PRACTICAL LESSONS FROM THE
LOMA PRIETA EARTHQUAKE

Report from a Symposium Sponsored by the
Geotechnical Board and the Board on Natural Disasters
of the National Research Council

Symposium Held in Conjunction with the
Earthquake Engineering Research Institute

Geotechnical Board
Commission on Engineering and Technical Systems

Board on Natural Disasters
Commission on Engineering and Technical Systems
and
Commission on Geosciences, Environment, and Resources

National Research Council

D1289047

NATIONAL ACADEMY PRESS
Washington, D.C.   1994

NOTICE: The project that is the subject of this report was approved by the Governing Board of the National Research Council, whose members are drawn from the councils of the National Academy of Sciences, the National Academy of Engineering, and the Institute of Medicine. The members of the panel responsible for this report were chosen for their special expertise and with regard for appropriate balance between government, industry, and academia.

This report has been reviewed by a group other than the authors according to procedures approved by a Report Review Committee consisting of members of the National Academy of Sciences, the National Academy of Engineering, and the Institute of Medicine.

The National Academy of Sciences is a private, nonprofit, self-perpetuating society of distinguished scholars engaged in scientific and engineering research, dedicated to the furtherance of science and technology and to their use for the general welfare. On the authority of the charter granted to it by the Congress in 1863, the Academy has a mandate that requires it to advise the federal government on scientific and technical matters. Dr. Bruce M. Alberts is president of the National Academy of Sciences.

The National Academy of Engineering was established in 1964, under the charter of the National Academy of Sciences, as a parallel organization of outstanding engineers. It is autonomous in its administration and in the selection of its members, sharing with the National Academy of Sciences the responsibility for advising the federal government. The National Academy of Engineering also sponsors engineering programs aimed at meeting national needs, encourages education and research, and recognizes the superior achievements of engineers. Dr. Robert M. White is president of the National Academy of Engineering.

The Institute of Medicine was established in 1970 by the National Academy of Sciences to secure the services of eminent members of the appropriate professions in the examination of policy matters pertaining to the health of the public. The Institute acts under the responsibility given to the National Academy of Sciences by its congressional charter to be an advisor to the federal government and, upon its own initiative, to identify issues of medical care, research, and education. Dr. Kenneth I. Shine is president of the Institute of Medicine.

The National Research Council was organized by the National Academy of Sciences in 1916 to associate the broad community of science and technology with the Academy's purposes of furthering knowledge and advising the federal government. Functioning in accordance with general policies determined by the Academy, the Council has become the principal operating agency of both the National Academy of Sciences and the National Academy of Engineering in providing services to the government, the public, and the scientific and engineering communities. The Council is administered jointly by both Academies and the Institute of Medicine. Dr. Bruce M. Alberts and Dr. Robert M. White are chairman and vice-chairman, respectively, of the National Research Council.

SPONSORS: The U.S. Department of the Interior, the U.S. Geological Survey, under Agreement No. 1434-92-A-1091, provided support to the National Research Council for this project. The National Science Foundation, Federal Emergency Management Agency, and National Institute of Standards and Technology provided support for the symposium through the Earthquake Engineering Research Institute.

Library of Congress Catalog Card Number 94-66357
International Standard Book Number 0-309-05030-8

B-275

iv

v

Preface

The Loma Prieta earthquake struck the San Francisco area on October 17, 1989, causing the loss of 63 lives and $10 billion of damage. As the results of the research, conducted in response to the earthquake, became known over the following three years, the U.S. Geological Survey, the sponsor of much of the research, approached the National Research Council (NRC) about how the results of the Loma Prieta earthquake research could be applied to other earthquake-prone areas of the country.

The NRC's Geotechnical Board and Board on Natural Disasters formed a committee, under the auspices of the NRC, to plan a major symposium on lessons learned from research and related activities conducted on the Loma Prieta earthquake. The Committee on Practical Lessons from the Loma Prieta Earthquake, chaired by Lloyd Cluff, accepted the responsibility to develop the symposium agenda, invite the speakers, review the keynote papers, and coordinate activities with the Earthquake Engineering Research Institute, which was responsible for managing the symposium and inviting symposium participants. The Symposium on Practical Lessons from the Loma Prieta Earthquake took place in San Francisco on March 22-23, 1993, and was attended by over 400 individuals.

The committee also agreed to develop these proceedings, which consist of six keynote papers solicited by the committee for the major sessions of the symposium: Geotechnical; Buildings; Emergency Preparedness and Response; Lifelines; Highway Bridges; and Recovery, Mitigation, and Planning. Selected remarks by panels of discussants on the applicability to other areas of the country of each keynoter's lessons are included following each keynote paper. The report also contains the opening keynote presentation by L. Thomas Tobin, the Executive Director of the California Seismic Safety Commission.

Drawing on the keynote papers and discussions at the symposium, an overview chapter has been written by the committee to present its summary of the principal lessons learned from the Loma Prieta earthquake. The overview contains, in addition, recommendations the committee believes are appropriate to improve seismic safety and earthquake awareness in areas of the country vulnerable to earthquakes but not as well-prepared as California.

The Geotechnical Board and the Board on Natural Disasters wish to thank the Earthquake Engineering Research Institute for its cooperation in planning and conducting the symposium. The boards also acknowledge the generous contributions of time and thought donated by all the speakers, discussants, and participants at the symposium.

Contents

APPENDIXES

PRACTICAL LESSONS FROM THE
LOMA PRIETA EARTHQUAKE

Overview: Lessons and Recommendations from the Committee for the Symposium on Practical Lessons from the Loma Prieta Earthquake

The moderately large (7.1 on the Richter Scale) Loma Prieta earthquake of October 17, 1989, took 63 lives, cost $10 billion, and damaged more than 27,000 structures. It resulted from a slip along a 25-mile segment of the San Andreas fault where it traverses the Santa Cruz Mountains, approximately 60 miles south of San Francisco and Oakland (Figure 1). Because the fault ruptured bilaterally, propagating north and south simultaneously, the duration of shaking was surprisingly short, about 10 seconds.

Within minutes of its occurrence, it was evident that the Loma Prieta earthquake would become an important case study for all interested in earthquake hazard assessment and risk reduction. Because it presented such an obvious opportunity to learn more about earthquake hazards and ways to mitigate their effects, the U.S. Geological Survey (USGS) asked the National Research Council (NRC) to convene a Symposium on Practical Lessons from the Loma Prieta Earthquake and to issue a report on the lessons learned from the research conducted on the earthquake. The symposium, conducted in conjunction with the Earthquake Engineering Research Institute (EERI), was attended by over 400 individuals. It took place on March 22-23, 1993, in San Francisco, California. The National Science Foundation, the Federal Emergency Management Agency, and the National Institute of Standards and Technology joined the USGS as sponsoring agencies. The symposium consisted of keynote papers, panel discussions, and technical sessions.

This report is the result of the symposium. It contains the opening keynote address, six keynote papers, and highlights of the discussants' comments. This overview chapter, authored by the NRC committee, contains the principal les-

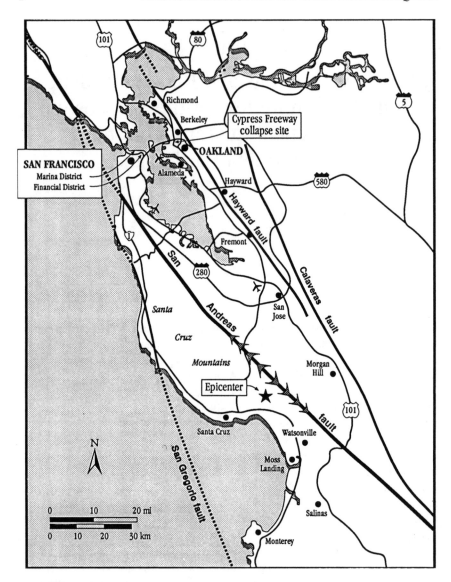

FIGURE 1   Map showing the location of the October 19, 1989, magnitude 7.1 Loma Prieta earthquake. Arrows show the extent and direction of fault rupture.

sons and recommendations drawn from the presentations made and discussions held at the symposium and the committee's collective opinions. In emphasizing the *practical* lessons of the Loma Prieta earthquake, the committee does not wish to downplay the importance of long-term research efforts, which have been the principal focus of other meetings. Rather, the primary objective of this conference was to respond to the more immediate concerns of user groups such as local and state governments, in terms of measures they might take at this time to better prepare for future earthquakes.

The Loma Prieta earthquake was the first opportunity in almost two decades to assess design and construction procedures used in this country to accommodate earthquake forces generated by a moderately large earthquake with widespread structural damage. It was the most damaging and costly earthquake to strike California since 1906. However, because of its distance from major population centers and the short duration of shaking, Loma Prieta was not a rigorous test of the performance of the Bay Area's built environment.

Many lessons were learned; some lessons reinforced prior understanding of earthquake vulnerability and design. Other lessons contradicted previous understandings. For example, an earthquake of the magnitude of Loma Prieta on the San Andreas fault would have been expected to be associated with significant surface faulting that did not occur. Some lessons identified social problems that challenged the efficacy of the emergency response efforts; some helped to identify new concepts with regard to earthquake vulnerability assessment and preparedness. In some cases, old lessons were re-learned.

Loma Prieta emphasized that:

- local geology and site soil conditions affect the severity of the shaking and the damage it causes to the built environment;
- older residential construction is vulnerable to failure that can cause extreme personal, social, and economic consequences;
- there is urgent need to improve the seismic resistance of unreinforced masonry and nonductile reinforced concrete building and bridge structures;
- many of the most successful mitigation efforts were the direct result of state legislation;
- practical, problem-focused research is needed to assess the appropriate repair and retrofit procedures for restoring and improving various building types;
- emergency response efforts are complicated by existing social conditions, such as homelessness; and
- preparedness pays.

Why did some sections of the Cypress Street viaduct collapse and not others? What were the structural characteristics that caused some older buildings to remain undamaged while nearby newer buildings sustained damage? Why did nearby buildings of similar type and age perform differently? Was it the design details? Was it the lack of adequate construction supervision to ensure proper construction? Was it the different site geologic and soil conditions? Was it a combination of these and other factors? How did retrofitted bridges and buildings perform? Could relief efforts have been more effective in ethnic neighborhoods? How does one manage droves of volunteers? These and other vital questions call for an urgent examination to seek explanations and solutions through multidisciplinary cooperation.

The purpose of this symposium was to understand how to apply the legacy of knowledge left by the Loma Prieta earthquake to reduce the impacts of future earthquakes throughout the United States. To accomplish this task, the symposium committee and participants attempted to understand and translate this knowledge into practical lessons and recommendations that can be implemented by the appropriate government agencies, designers, builders, care-givers, and others responsible for people's well-being and quality of life.

In light of the presentations and discussions at the symposium, the following lessons and recommendations are offered by the NRC committee in the hope that these short statements will stimulate the reader to read further and to take appropriate action. The committee recognizes that within earthquake-prone regions of the United States there are varying levels of seismic risk. These lessons and recommendations do not distinguish between these risk levels but are directed, in general, to areas where risk is sufficient for seismic safety to be a priority.

One of the primary purposes of the conference was to emphasize the interdisciplinary nature of seismic hazard mitigation; the lessons and recommendations demonstrate this well. The forty lessons and recommendations listed below have been ordered so that the overall flow of thought is from general observations, to the earth sciences, to geotechnical and structural engineering, and finally to planning and emergency response. Nevertheless, the committee points out the cross-cutting nature of most of the lessons and recommendations, and it urges readers to review the entire list regardless of their own individual specialties.

GENERAL OBSERVATIONS

Lesson 1: Investments made in earthquake preparedness and hazard and risk mitigation paid off. It was apparent from the papers presented at the symposium that the Loma Prieta earthquake demonstrated that the San Francisco Bay Area has made progress toward improving the ability to minimize damage and cope with destructive earthquakes. However, Loma Prieta was a moderately large earthquake that occurred a significant distance from major population centers; therefore it was not a rigorous test.

Recommendation: Rather than creating over-confidence and complacency, the Loma Prieta earthquake must serve as a stern warning to residents of the Bay Area and other earthquake-prone areas throughout the United States about future earthquakes, some of which will be larger, shake longer, and be closer to major population centers.

Lesson 2: Many earthquake professionals knew that the San Francisco Marina District was prone to liquefaction, that unreinforced masonry buildings were vulnerable to earthquakes, and that existing older concrete structures probably lacked adequate steel reinforcement and connection details vital to resist strong shaking. However, government and business leaders expressed great surprise that these professionals knew so much, but so little was done. The failures stemming from the Loma Prieta earthquake reinforce the need for a communication program to close the knowledge gap among researchers, practicing professionals, decision makers, and the public in a prioritized and systematic way.

Recommendation: Earthquake professionals must work harder to close the gap between what is known and what is used, between researchers and practitioners. They must increase their emphasis on integrating seismic risk concerns and their existing knowledge into the mainstream activities of both government and business. There is a strong need for advocacy; narrowing the gap by enacting new policy depends on earthquake professionals having a commitment to do so. Practicing professionals must adequately inform policy makers of the risks and the costs and benefits of various strategies; they must make strong recommendations for earthquake hazard and risk mitigation, including evaluation, assessment and enforcement measures to assure effective implementation.

The National Earthquake Hazard Reduction Program must emphasize the application of existing knowledge and the development of new knowledge and must provide incentives for risk reduction activities.

EARTH SCIENCES

Lesson 3: In the first few hours following the Loma Prieta earthquake, uneven and in some cases inappropriate emergency responses resulted from the inability of emergency-response decision makers to know where the heaviest shaking and greatest damage actually occurred and from undue dependence on news media reports.

Recommendation: In the principal earthquake-prone areas of the country, federal and state governments, in particular, should accelerate efforts to develop on-line, real-time seismic monitoring systems capable of producing generalized maps of the intensity of heavy ground shaking and the relative vulnerability of the built environment within a few minutes of a major earthquake. Such systems are achievable with currently available technology, and prototype systems are currently under development.

Lesson 4: Even in the San Francisco Bay Area, where the seismic geology is well mapped and seemingly well understood, surprises occurred in the way in which the San Andreas fault zone ruptured and in the resulting geographic patterns of seismic shaking.

Recommendation: Federal, state, and local governments should continue to support the preparation of seismic zoning maps and emergency response plans, recognizing that geologic knowledge is necessarily limited, that local earthquakes of the historical past will not be exactly like those of the future, and that surprises will continue to occur. Continued studies by researchers on the mechanics of the earthquake rupture process and the prediction of destructive effects are essential; maps should be updated as soon as more reliable data become available.

Lesson 5: Damaging intensities of seismic shaking occurred at considerable distances from the epicenter of this moderately large earthquake. Strong crustal reflections resulting from regional geologic structure may have been of particular importance at these large distances.

Recommendation: In preparing emergency response plans, communities must recognize, and scientists must emphasize, that seismic hazard is by no means limited to local faults, nor will significant damage be caused only by "the big one." Data on seismic velocity structure and regional geologic structure, including crustal thickness, are significant in assessing the potential for local damaging ground motions from distant events, and should be assessed.

Lesson 6: Surprising variations in the intensity and nature of strong ground motion were seen in the close-in epicentral area as a function of proximity to the causative fault, rupture direction, azimuth to the epicenter, frequency of ground vibration, and other factors.

Recommendation: Seismologists and geologists must increase their efforts to understand and quantify the nature of strong ground shaking very close to highly active faults, and engineers must give special attention to the placement and design of structures in these areas.

GEOTECHNICAL AND STRUCTURAL ENGINEERING

Lesson 7: As has been observed in many earthquakes, and was dramatically re-emphasized by the Loma Prieta earthquake, the intensity of seismic shaking is critically dependent upon the nature of the local soils and shallow geologic structures. Major differences in intensities of shaking were observed over distances of only a few hundred feet in areas where soil conditions changed rapidly, such as near the edges of old stream channels or along the boundaries of filled ground. Most of these features were well delineated on pre-existing maps, but the impact of the possible hazards had not been fully incorporated into land-use plans or building codes.

Recommendation: Federal, state, and local governments should make an expanded effort in mitigation planning to use current knowledge of the effects of prior shaking and local soil and geologic conditions in major earthquake-prone metropolitan areas of the country. Microzonation should be used for new construction and for rehabilitation efforts. Furthermore, agencies must address how seismic hazard maps can be used effectively in land-use planning, design, and in administrative practices.

Lesson 8: Geologic maps prepared for the purpose of identifying potential areas of liquefaction proved accurate in defining the locations of major occurrences of liquefaction and lateral spreading. Where compaction, restraining walls, piles, vibroflotation, or other remedial methods were implemented, they significantly reduced the liquefaction susceptibility, although significant liquefaction damage occurred in adjacent areas of unimproved ground.

Recommendation: In earthquake-prone areas subject to potential liquefaction, practicing professionals should use regional maps of potential areas of liquefaction as guidelines to indicate where detailed site-specific studies need to be performed to characterize accurately geologic, soil, and groundwater conditions. When facilities must be built in areas identified as being susceptible to liquefaction, measures must be taken to ensure that ground improvement techniques are used to minimize earthquake damage.

Lesson 9: Loma Prieta caused about 4,000 landslides. Throughout the affected area, large, deep-seated ancient landslides were reactivated, many of which had been specifically delineated and recognized as potentially active prior to the earthquake. Nevertheless, little effective land-use planning and few mitigation measures had been carried out. Fortunately, the extent and impact of landsliding were moderated by the effects of a four-year drought, as well as by the moderate size of the earthquake, the short duration of shaking, and the distance of the energy release from large populations.

Recommendation: Appropriate government agencies should continue to develop maps of landslide potential. These maps should become an integral part of land-use planning, and the information should be communicated to developers and building owners. Communities having high landslide potential should not be complacent when reviewing the lessons from the Loma Prieta earthquake—what appears to have been successful performance could have turned out to have unacceptable consequences if the earthquake had occurred closer to cities, the shaking had lasted longer, or the ground had been highly saturated.

Lesson 10: Earth and rockfill dams within the epicentral region built to California's modern standards of construction performed well. Some old dams suffered significant damage. Fortunately, because of the drought that had plagued California for several years, very little water was retained in most of the reservoirs at the time of the earthquake.

Recommendation: Responsible government agencies should ensure that all dams are designed and built to a high standard. Older dams should be reexamined and retrofitted to improve stability and safety where deficiencies are found.

Lesson 11: Where specific geotechnical engineering measures had been taken to compact major artificial fills throughout the Bay Area, these fills performed well.

Recommendation: Placement and compaction using appropriate geotechnical engineering should be mandatory wherever filled ground is to be used in earthquake-prone areas. Communities having existing fills should evaluate the ability of the fills to withstand stronger shaking than that caused by the Loma Prieta earthquake, and should take remediation steps, if necessary.

Lesson 12: With the appropriate application of existing knowledge and with more attention to detail, practicing professionals licensed to site, design, and retrofit buildings and lifelines could have significantly reduced the loss of life and the damage resulting in loss of function.

Recommendation: Professional and trade organizations should place a priority on improving the knowledge of practicing designers, engineers, geotechnical professionals, and building contractors. These organizations should develop training programs for practicing professionals who should be provided incentives to upgrade their knowledge. Frequent training is needed to keep their skills and methods up to date so they can effectively apply the knowledge learned from past earthquakes. Building owners need to be encouraged to hire professionals who have demonstrated competence in addressing earthquake hazards.

Lesson 13: In some cases, bridges and buildings that *had* been retrofitted to resist earthquakes sustained structural damage. In other cases, many bridges and buildings that *had not* been retrofitted sustained little or no damage. (See "Note on Supporting Sources" at the end of this chapter.)

Recommendation: To assess the appropriate retrofit procedures for improving various bridge and building types, practicing professionals need to know more about bridge and building performance during earthquakes. Professionals and researchers need to focus efforts to correlate actual re-

sponse with predicted response using realistic earthquake scenarios. This requires studies of the expected earthquake response of specific bridges and buildings. Both retrofitted and non-retrofitted structures should be thoroughly instrumented to allow direct measurements of their actual response. Appropriate state and local government agencies should take the lead in encouraging the instrumentation of critical bridges and buildings in metropolitan areas that are close to sources of likely future large earthquakes, such as Seattle, Portland, San Francisco, Los Angeles, San Diego, Charleston (South Carolina), the Wasatch front region in Utah, and the New Madrid seismic zone in Missouri.

Lesson 14: The good performance of "hinge restrainers" is testimony that highway bridge structures can be economically retrofitted to enhance seismic performance. This simple, relatively inexpensive, interim retrofit technique provided displacement control across expansion joints.

Recommendation: Transportation departments can obtain a significant improvement in seismic resistance in most simple span bridges or bridges having hinge seats by installing cable restrainers. Although not a complete solution, they are an interim "quick fix" that can provide a measure of protection at minimal cost.

Lesson 15: Many owners and managers of buildings that were built to the Uniform Building Code or had been seismically upgraded were surprised when their buildings experienced significant damage. In some cases, the owners were not able to reoccupy their buildings for some time after the earthquake and could not understand why.

Recommendation: Earthquake professionals must educate owners of buildings to understand that building code compliance will not and was never intended to make buildings earthquake proof. The codes and most seismic strengthening of buildings are intended to protect life and to minimize catastrophic damage. A modern building that meets current codes or an older one that has undergone retrofitting can experience damage that could cause it to be closed for what may be an unacceptable period of time. Owners can make informed decisions on investments in greater earthquake resistance to achieve a superior level of building performance during earthquakes and of post-earthquake integrity. Engineering estimates can indicate what level of performance owners can expect at what cost. Practicing professionals need to develop refined techniques, consistent with the historical behavior of structures, to make such estimates.

Lesson 16: There were numerous structures, ranging from residential buildings to multistory facilities, that were seriously damaged in the Loma Prieta earthquake because of design or construction errors that should have been found during the building inspection process. Local building departments rarely have the necessary resources to accomplish this fundamental step in the construction process, and design professionals often shy away from providing the needed oversight. (See "Note on Supporting Sources" at the end of this chapter.)

Recommendation: In regions of the United States where seismic safety must be a priority, local governments should insist on adequate inspection and enforcement of construction regulations and standards. Similarly, regulators need to demand attention to detail, as well as procedures that result in construction quality. Design professionals should be required to inspect and accept the work. Educational courses should be mandatory to provide building inspectors with up-to-date principles of seismic design that cover the majority of structures. Local governments should provide qualified and properly trained building inspectors who have adequate resources to confirm that all the seismic elements are properly installed.

Lesson 17: Unreinforced masonry buildings are not only dangerous to building occupants but can be equally dangerous to adjacent buildings and people on sidewalks and in the streets, due to falling debris.

Recommendation: Local governments should create mandatory procedures or programs to reduce the seismic hazards associated with unreinforced masonry buildings in all cities that may be exposed to damaging earthquakes.

Lesson 18: For residential structures, simple retrofits made prior to the earthquake greatly reduced losses. The simple and inexpensive procedures of strengthening cripple-walls, bolting residences to their foundations, and anchoring water heaters have a very high benefit/cost ratio.

Recommendation: Homeowners should be encouraged to perform simple retrofits. Earthquake-prone communities desiring guidance for establishing formal procedures or recommendations for homeowners should review the California Seismic Safety Commission's publication, The Homeowner's Guide to Earthquake Safety. State law requires home sellers to disclose to buyers all earthquake weaknesses; real estate agents are required to distribute the publication.

Lesson 19: Loma Prieta re-emphasized the need to understand and deal effectively with "adjacency hazards." These include debris or building components falling from one building and damaging the adjacent building, adjacent buildings pounding each other, and the interaction of differing heights of adjacent buildings. All of these adjacency hazards were observed throughout the area affected by the Loma Prieta earthquake. There are currently no codes or ordinances that deal with these hazards.

Recommendation: Local governments should encourage cooperation between adjacent building owners when adjacency hazards are identified, so that life-threatening hazards and economic impacts can be minimized.

Lesson 20: Experience during the Loma Prieta earthquake confirmed the inherent ruggedness of modern steel natural gas transmission pipelines and distribution mains. Damage to gas mains was generally limited to old cast iron pipe in areas that experienced ground failure due to liquefaction, landslides, and extreme settlement of natural and man-made fills. This behavior is consistent with observations from past earthquakes worldwide.

Recommendation: Gas pipeline vulnerability studies must include the assessment of areas of potential ground failure, including liquefaction, differential settlement, and surface faulting. These hazards should be mitigated by replacing cast iron pipes with modern steel ones and by designing plans for rerouting or rapid repair.

Lesson 21: Well-designed water and wastewater-treatment facilities experienced little structural damage; however, non-structural components were prone to damage due to sloshing effects.

Recommendation: Non-structural components of water and wastewater-treatment facilities should be designed using the same principles as those for water storage tanks.

Lesson 22: As a moderately large earthquake of short duration, Loma Prieta resulted in only a few cycles of strong shaking and demonstrated that older bridges on soft soil are vulnerable to even this limited level of ground motion. Extensive damage to similar transportation structures elsewhere in the United States must be expected.

Recommendation: Earthquake-prone communities throughout the United States should examine the vulnerabilities of their bridges. Screening procedures to set priorities, and retrofit methods are available that have immediate application in other areas. These are the result of extensive research funded by the California Department of Transportation, the Federal Highway Administration, and the National Science Foundation. They are applicable throughout the United States without the need to "reinvent the wheel."

Lesson 23: High-voltage (500-kilovolt) substations were the least seismically resistant element in the electric utility bulk power transmission system. Power system failures were largely due to older-vintage, live-tank circuit breakers; no dead-tank circuit breakers failed.

Recommendation: Utilities should replace older-vintage, live-tank circuit breakers with dead-tank circuit breakers or new live-tank breakers that have been shake-table tested.

Lesson 24: The results of earthquake research have not been effectively translated to many potential users of the research findings in the Bay Area. Earthquake research often is viewed as too narrowly focused, not particularly useful, or inadequately communicated to practitioners.

Recommendation: There must be a balance between basic research and problem-focused research. To improve the transfer of research technology to the research-user community, strong partnerships should be formed among researchers, design professionals, industry, and federal, state and local government agencies. One of the most effective ways to initiate these partnerships is to involve the practicing professionals and the end-users of research in problem-focused earthquake research programs. This involvement should include the research users in the planning and scoping activities, as well as in decision-making activities during implementation of the research.

Lesson 25: The disruption of water pipelines resulted in the rapid loss of water supplies stored in elevated water tanks, depleting water vital for fire fighting.

Recommendation: Local agencies and the appropriate public utilities should install "smart" valves in the system to shut off the loss of water and protect the water supply when a break in the system has occurred. These shut-off valves should be sited on the upstream side of potential ground failure areas. Automatic shut-off valves that are activated by strong shaking are not recommended, because they often shut off the water supply to pipeline systems that have not experienced damage and the water may be needed to fight fires.

PLANNING AND EMERGENCY RESPONSE

Lesson 26: During the response period, there was an urgent need for better coordination among the various levels of government and private sector businesses. Those organizations that had developed and tested realistic earthquake planning scenarios prior to the Loma Prieta earthquake were better prepared than those that had not. Good emergency response and recovery plans are essential to facilitate coordination and to enable quick response and full recovery following even a moderate earthquake.

Recommendation: Local governments, with assistance from state or federal agencies, utilities, or other organizations, should develop realistic earthquake scenarios to evaluate the vulnerability of their communities (including lifeline systems, businesses, neighborhoods, and other community investments), to test emergency response plans and communication links, and to gain insight for recovery plans. Large organizations, businesses, and utility companies should conduct earthquake exercises periodically to practice emergency response and communications. These organizations should use the resulting understanding of earthquake vulnerabilities to implement realistic, priority-driven earthquake risk reduction and risk management programs.

Lesson 27: Power outages in downtown San Francisco lasted several days following the earthquake due to the need for time-consuming inspections of major buildings for gas leaks and ignition sources prior to energizing the downtown power grid. This was the largest single source of business interruption resulting from the Loma Prieta earthquake.

Recommendation: Local agencies and the appropriate public utilities should develop building inspection procedures and train personnel to facilitate the building inspections for gas leaks.

Lesson 28: Existing federal disaster recovery policy is written in a way that inhibits upgrading of hazardous structures. Disaster recovery laws are biased toward returning to pre-earthquake conditions, even when those conditions represent high earthquake risk.

Recommendation: The federal government should maintain flexibility in recovery policy to react to changed conditions and to reflect the need for seismic hazard mitigation. "Exact replacement" is unsound public policy. With such flexibility, future opportunities to improve seismic safety are enhanced, and damaged or collapsed facilities can be replaced with improved seismic resistance. Appropriate government agencies and professional and trade organizations should develop guidelines and standards to guide earthquake repair in a way that provides for a variety of performance levels that are based on the facilities' intended function and the level of safety desired. Federal procedures for awarding earthquake recovery funds to state and local governments should require that the federal contribution be used to restore the stricken community to a functioning, viable community that has improved seismic safety.

Lesson 29: Pre-existing social problems such as homelessness, housing shortages, tight government budgets, land-use disputes, and inadequate lifelines will be made worse immediately after a destructive earthquake.

Recommendation: Community leaders should incorporate the social problems that currently exist in their communities into their emergency planning. The leaders should consider how such social problems are likely to be exacerbated during the emergency response and recovery period. Community agencies need to develop policies to address these issues and to develop the capacity to innovate. For example, volunteers can be trained to provide a support service to police and fire units as is done by the Los Angeles Fire Department.

Lesson 30: There was an outpouring of unselfish concern for the welfare of others, even though the massive response often created widespread confusion and coordination problems. People were adaptive and flexible. Destructive earthquakes invariably produce unexpected challenges for responders whose effective action is dependent on their education and training and on their willingness to develop and help implement innovative solutions on the spot. When well informed about earthquake hazards, people significantly improve their chances of safety. For example, the Oakland Fire Department has 5,000 citizen volunteers who have completed training modules and are now a support service to police and fire units. These individuals can lead the emergent volunteers and help to manage them so they are a resource and not a liability.

Recommendation: Local governments must show a commitment to earthquake hazard mitigation and preparedness efforts so that the populace's resiliency and ability to function in response to the earthquake is improved.

Lesson 31: Rushed post-earthquake inspections of damaged buildings resulted in some inaccurate and emotional assessments that led to inappropriate actions (for example, demolition) in the spirit of protecting public safety, which caused extreme financial impact to owners and renters.

Recommendation: Local governments should develop and implement strategies, procedures, and training for post-earthquake inspections. These should be communicated prior to the event to avoid costly mistakes. New, realistic guidelines for emergency demolition are needed.

Lesson 32: Within the Bay Area's communication systems, sufficient slack in fiber-optic cable assisted in minimizing failure and service disruptions, even in areas where large differential displacements occurred. For example, the fiber-optic cable attached to the Bay Bridge was not disrupted, even though one of the upper spans of the bridge collapsed during the earthquake. Although the cable stretched, there was sufficient slack to accommodate the damage to the bridge.

Recommendation: Appropriate utilities should incorporate cable slack as a design requirement in order to minimize failure of the fiber-optic communication systems.

Lesson 33: In many culturally diverse communities in the Bay Area, the languages and customs were not understood by relief workers after the Loma Prieta earthquake. As a consequence, actions by relief workers were not as effective as they might otherwise have been.

Recommendation: Relief agencies must be knowledgeable about the ethnic composition of at-risk areas and must be prepared to train relief workers in the language and customs of diverse populations, so that organized efforts to provide assistance to earthquake victims will be effective.

Lesson 34: The demolition or preservation of damaged historic buildings was and remains a bitterly fought issue in many Bay Area communities. Historic buildings can be a community asset, but they carry with them certain costly responsibilities for the owner.

Recommendation: *Communities need to address the issue of historic buildings through a well-considered, long-term plan, prior to a destructive earthquake. Government entities and private historic preservation organizations should consider providing incentives to strengthen chosen historic buildings and repair them after damaging earthquakes.*

Lesson 35: More-effective education of the public is needed about risks that are related to natural gas leakage following earthquakes. Despite efforts by California natural gas companies to educate the public, almost all of the 156,000 customer gas shut-offs following the Loma Prieta earthquake were unnecessary. They were mostly initiated by poor advice from the media immediately after the earthquake.

Recommendation: *Public education needs to be repeated whenever there is an opportunity. Pre-recorded audio and video tapes should be furnished to the media by natural gas companies, and notices should be inserted in utility bills periodically. Emergency response training for all media reporters should be encouraged.*

Lesson 36: Disagreements over damage estimates, the cost of repairs, and the level of expected performance were frustrating and caused restoration delays. For example, a peer-review panel was selected by the California Department of Transportation immediately after the earthquake; however, this panel did not convene until March 1990. Following its review, some of the repairs to transportation structures, which had already begun, had to be halted or abandoned and much of the design redone, resulting in considerable delay and wasted effort and money.

Recommendation: *State and local governments and departments of transportation should select and contract with seismic peer-review panels in anticipation of future earthquakes. The panels should be charged with the review of designs for new facilities, as well as retrofit concepts and implementation methods for mitigation measures for old facilities. A clear set of instructions should be established for these panels ahead of time so that both the panels and the owners know what to expect and know the limitations on each party.*

Lesson 37: Many of the most successful mitigation efforts were the direct result of state legislation, such as the California Field Act to protect schools and the California Hospital Act to protect hospitals. Nevertheless, it appears that many local governments in California and elsewhere do not adopt seismic safety requirements to protect the local population.

Recommendation: State governments should provide legislative incentives for local governments to adopt ordinances and to work cooperatively with adjacent and regional governments and with state and federal agencies to implement seismic safety improvements.

Lesson 38: The low-income population of the Bay Area was most seriously impacted, because these individuals typically occupy old, seismically weak buildings that have not been adequately maintained. Many of these buildings were so badly damaged that they had to be evacuated. This increased the homeless population during the response phase, thus adding an increased burden to already over-burdened local governments, as well as to mass-care and shelter providers such as the Red Cross. In 1993, some of these buildings still stand empty, waiting for decisions and actions. Many are in no worse condition than they were before the earthquake. Advocates for the homeless believe that seismic upgrades will increase rents and make the buildings unavailable to the poor. Regulators responsible for seismic safety require the buildings to be upgraded. Some building owners simply choose to demolish their buildings rather than do the upgrade.

Recommendation: Communities must identify seismically weak buildings, especially in poor neighborhoods, and solutions should be developed prior to a destructive earthquake. Guidelines for evacuation and re-use are needed. Realistic social and economic incentives are necessary to motivate owners to bring older buildings up to acceptable standards. Earthquake professionals must take note of the factors that motivate building owners and business leaders: improving return on investments, lowering expenses, curtailing losses, and avoiding liability.

Lesson 39: The fact that the Loma Prieta earthquake occurred as the third World Series game between San Francisco and Oakland was starting resulted in extensive international media coverage and increased earthquake awareness.

Recommendation: Local governments and other public and private entities should be prepared to capitalize on the greater awareness of earthquakes that occurs immediately after an event. If agencies have well thought out plans prior to an earthquake, such plans are more likely to be adopted following the event. Communities respond more readily to preparedness and mitigation advice when the citizens understand that they are also at risk.

Lesson 40: Recovery from destructive earthquakes is expensive for everyone. Federal disaster aid pays for only a small portion of the recovery expenses. The damage losses from the Loma Prieta earthquake are estimated at about $10 billion. The federal contribution, through a variety of federal programs, will amount to approximately $2.5 billion. The state of California levied a 1/4-cent sales tax increase for 13 months and raised about $800 million. Insurance payments totaled about $1 billion. The remaining $5.7 billion—more than 50 percent of the losses—must be accommodated by local governments, businesses, and private individuals.

Recommendation: Earthquake hazard and risk mitigation and education efforts must be accelerated in earthquake-prone areas. The Loma Prieta earthquake dramatically demonstrated that an investment of a few percent of the $10 billion would have paid off many-fold in reducing the earthquake losses (death, injury, and economic impact). Citizens must do everything possible to convince decision makers of the value of investing in pre-earthquake mitigative measures.

NOTE ON SUPPORTING SOURCES

See the 1989 *Hearings on the Loma Prieta Earthquake* conducted by the California Seismic Safety Council in which numerous examples are described to support Lessons 13 and 16. Other references include: California Seismic Safety Commission, 1991, *Loma Prieta's Call to Action*, 97 pp., Sacramento; Lew, H.S. (Ed), 1989, *Performance of Structures During the Loma Prieta Earthquake of October 17, 1989*, U.S. Department of Commerce, National Institute of Standards and Technology, Special Publications #778, 175 pp.

1

Legacy of the Loma Prieta Earthquake: Challenges to Other Communities

L. Thomas Tobin

INTRODUCTION

The National Research Council and Earthquake Engineering Research Institute are to be congratulated for conceiving and sponsoring a symposium on practical lessons from an earthquake. Far too often, we treat earthquakes as abstract laboratories of scientific curiosity, as events that raise questions and provide opportunities for further studies. Focusing on practical lessons is a start, but our object should be, as with all lessons, to translate them into action that reduces risk.

The charge to each of us attending this symposium is to determine how we can apply our knowledge to reduce risk. The Loma Prieta earthquake and other earthquakes before and since have left us a legacy of information. I think we are all generally aware that we possess the knowledge to reduce earthquake risk across the nation. If a general reduction of risk is to be our legacy, however, we must embed our information, knowledge, and understanding in public policy.

We are doing relatively well in synthesizing information and knowledge into understanding. However, while each day brings more understanding of earthquakes, soil types, materials, and designs, policies that implement this increased understanding are not advancing as fast. We all must learn how public policy is created. We must expand our scope of collegiality and become active policy advocates. The future cannot afford our comfortable and exclusive association with other engineers and scientists. We must add building officials, city councils, boards of supervisors, and state legislators as colleagues to a broadened endeavor. We know how to build a bridge to resist earthquakes, now we have to learn how to build a bridge to policy makers to ensure that only safe bridges are built. The current strategy employed by the National Earthquake Hazard Reduction Program (NEHRP) is to support research. Others, working in state and local government, are expected to implement the results of their research through concrete actions. Unfortunately, this approach does not lead to substantial progress in coping with our vulnerabilities to earthquakes. Trickle-down mitigation works no better than trickle-down economics.

THE NEED TO REACH OUT

The Loma Prieta earthquake provided a perspective different from earthquakes in Mexico City, Armenia, Ferndale, Landers, and Whittier. After observing damage in my home town and talking to families of victims, I was ashamed that we had not fully used what we knew. We are all culpable for failing to use our knowledge to effect change. We spend too little time using what we know to change public policy. I recognize that many of you fully use available knowledge in your practice, but most of us expect others to change public policy. This is the fundamental flaw in our efforts; ignoring public policy is an abdication of our responsibility. We are failing in this, our ultimate job, by not swaying those whose decisions affect public and corporate policy.

We also are not reaching our colleagues. Look around this room. You do not see 99 percent of the 560 building officials in California, 99 percent of the practicing civil engineers, architects, or engineering geologists. How do we reach these people? Who is responsible for reaching them? It's not the other person; it's you and me. It's our responsibility to use what we know to foster widespread professional training.

Our challenge is to reach those who are not here, to teach other professionals, and to sway policy makers. You can press these ideas in your professional societies; before boards of registration; and by serving on advisory committees, planning commissions, and city councils. You can demand mandatory professional education. You can influence your elected representatives. You can use your knowledge to see to it that earthquake risk is reduced.

THE NEED FOR SUSTAINED ADVOCACY

The Loma Prieta earthquake reinforced common-sense wisdom that there is a "window of opportunity" for seismic safety advocacy in the aftermath of an earthquake. One must be ready with proposals for state and local legislation and private-sector clients. The earthquake left a legacy of new and enhanced public and private programs begun during the window of opportunity:

• The California legislature considered over 300 seismic bills. They created a new seismic hazard mapping program, a prepaid residential recovery fund, and an earthquake deficiency disclosure requirement for residences and certain commercial buildings; placed a $300 million bond measure on the ballot (which passed when other bond measures failed); and required the California Department of Transportation—CALTRANS—to retrofit existing vulnerable bridges.

• The cities of San Francisco, Los Angeles, Berkeley, and Oakland passed general obligation bond measures to retrofit certain buildings and improve emergency response capabilities. The state and Santa Cruz County temporarily raised sales taxes to pay for repair and retrofit.

• Statewide, many cities and counties accelerated their efforts to comply with the state law that requires the identification of unreinforced masonry buildings, the notification of owners, and the adoption of mitigation programs. Only 32 jurisdictions had adopted unreinforced masonry (URM) mitigation programs on October 17, 1989; nine months later the total was 154. Today 286 or 78 percent of the affected jurisdictions are in compliance with the law. Over half of the jurisdictions have programs in place that require owners to retrofit or raze their buildings within a specified time.

• The cities of Alameda, Palo Alto, and Santa Clara have entered into a mutual-aid agreement with the southern cities of Burbank, Glendale, and Pasadena to share personnel and equipment needed for the repair of municipal utilities. According to the agreement, the cities needing assistance would pay for personnel, transportation, and materials sent by the responding cities.

• The state Office of Emergency Services has established a heavy search and rescue capability and is now installing the "OASIS" satellite communications system, which links key state agencies with county emergency operation centers.

• The private sector's response also has been impressive. Businesses in southern California and in the Bay Area have acted to retrofit their buildings, diversify their production, and prepare for their emergency response and business recovery.

California took advantage of the "window of opportunity" to make a number of long-term advances, because it had an existing advocacy program and a plan. Waiting for a window of opportunity is no substitute for ongoing, persistent policy advocacy. The window may not open very wide, and it will not remain open for long.

• Even in late 1989 and early 1990, before the dust of the Loma Prieta earthquake had settled, the economy and budget loomed as the most important issues in Sacramento. The legislature would not deal with a number of seismic issues, because their priority was recovery, not mitigation.

• Not every policy body will act while the "window of opportunity" is open. Nearly 100 cities and counties still have not adopted URM mitigation programs and are out of compliance with a state law passed three years before the Loma Prieta earthquake. Another 100 have either voluntary or notification-only programs, which, though nominally in compliance, are ineffective.

• Carrying out and sustaining advances is proving more difficult than creating new programs. An economy in recession all but closed the window in California by 1991, and now we are learning that policy gains are not necessarily permanent.

• Performance is not guaranteed. Even when programs are established and underway, there needs to be a constant push for performance and a demand for accountability. Following the Loma Prieta earthquake, a state agency decided that the state should lease buildings with a reasonable chance of withstanding earthquakes. They proposed a policy to that effect in 1990. Unfortunately, because of retirements in the ranks and a change in administration with new leadership, the proposal was lost in an institutional limbo for nearly two years.

The "political mood" related to the economy also has undermined a number of our gains. Private businesses and local and state governments are cutting back substantially; they cannot afford "too much" safety. We are experiencing an erosion of our advances.

• The California Residential Earthquake Recovery Fund was repealed. The opposition, driven largely by insurance industry political interests, was unable or unwilling to stop passage in 1990, but just two years later, the climate had changed, and fear that claims might exceed resources in the fund was greater than the concern for future earthquake victims. A majority of legislators voted for repeal.

• There is an ongoing effort to emasculate the 60-year-old Field Act. Until now, our schools could be counted on to protect children and provide emergency shelter. That may change.

• In virtually every local government, budget cuts are significantly reducing the ranks of professional emergency managers. We may be saving pennies for our future, but at what hidden price?

• The Seismic Safety Commission faces budget hearings in which there is the distinct possibility that funding for the Commission will be eliminated.

Thus, while the "window of opportunity" does exist, and while we should be ready to take advantage of it, it is far more important to build a sustained advocacy program. Otherwise, we may find ourselves waiting forever for the "right time."

Political and public support for earthquake risk reduction measures can be developed by making decision makers aware of the hazard; explaining the expected losses to the buildings, businesses, and functions they care about; and then explaining that there are technically and politically viable solutions and funding within existing administrative structures. The familiarity and trust that develops from a credible, long-term advocacy program is essential and even more important when the "window" is open. Sustained advocacy is essential to changing public policy and putting our legacy of knowledge to work.

IRREVERSIBLE CHANGE

If we fully used available knowledge, our public policies and key programs would be logical extensions of that knowledge. However, the Loma Prieta earthquake shows this is not the case. Consider public policy relating to recovery. The earthquake caused irreversible changes: personal losses, structural losses, cost increases, different options for injured persons, disrupted businesses, and shattered neighborhoods, and it created different responsibilities for state and local governments. Accepting change runs counter to the human inclination to return as rapidly as possible to "normal," to return to "the way it was." Unfortunately, once damaged deeply enough, the community will never be the same.

Changes are not caused just by the earthquake. Pre-existing problems such as housing shortages, run-down and underused commercial areas, and lack of governmental funding are revealed and made worse by an earthquake. Needs change as well. The business climate may have changed, the building stock may have deteriorated, demands on the existing infrastructure may have grown, and jobs may have moved elsewhere. New problems, such as homelessness, narcotics-related crime, and new land uses may have changed the community. The recovery effort, especially building and infrastructure repairs, should reflect these changes and anticipate the ensuing policy and land-use disputes.

Our national disaster recovery policy, however, is unresponsive to this reality. It is crafted in a way that inhibits change and is slow to resolve controversy. Recovery policies are biased toward returning to the status quo, toward replacing what was there. They do not provide flexibility to accommodate changed conditions and needs.

Not too far from here, work is progressing on the demolition of ramps connecting city streets with the Bay Bridge. Approximately $175 million of federal money is available to replace these damaged ramps. CALTRANS fears changing the alignment will jeopardize the funds, even though San Francisco and its traffic patterns have changed greatly since the ramps were built in the 1930s. The planning director for San Francisco, Lucien Blazej, was quoted as saying, "We should be planning ahead. Rebuilding what exists today does not address that issue. [Exact replacement] is bad public policy at every level."

Post-disaster mitigation funds are insufficient to repair buildings to resist future earthquakes. Moreover, disaster programs presuppose the existence of standards that allow a consistent way to evaluate damage and determine the cost of repair. We all know standard repair codes are not available and that a "single code" cannot provide the flexibility to achieve the performance level appropriate to the use of each repaired building.

Arguments between applicants for aid and funding agencies and between proponents of the various causes are supported by professionals. Their opinions are founded on different levels of expertise, investigation, and assumptions about performance. It is a Catch-22. Disagreements over damage estimates, cost of repairs, expected performance, and changed community needs cause delay and frustration. In the end, no one is well served. Judgment and flexibility, specific to the conditions of each building and disaster, are needed to foster recovery.

Failing to resolve disagreements in a timely and flexible fashion will delay recovery, increase costs, and enrich attorneys. A conflict resolution mechanism is needed. Peer panels could be used to resolve both factual and judgmental disagreements. Binding arbitration could be used to make timely decisions. Federal agencies could give state and local governments recovery funds as block grants, requiring only that the funds be used to return the disaster area to a functioning and viable community. We need repair guidelines and standards that provide for a variety of performance levels based on the building and its intended use. A legacy of the Loma Prieta earthquake is the knowledge that the federal disaster aid program should be changed.

ROME WASN'T BUILT IN A DAY

The Loma Prieta earthquake reminded us that recovery is a long-term effort that imposes a substantial financial burden on local governments and businesses.

• The city of Santa Cruz is experiencing the painfully slow process other cities should expect. Over 325,000 square feet of office and commercial space on Pacific Avenue was destroyed by the earthquake. About 200 businesses and 1,400 employees were displaced, and 45 buildings were damaged or destroyed. Economic losses were estimated at $100 million. Only within the last couple of months—more than three years after the quake—has the city completed the reconstruction of the utilities and the infrastructure. Now building owners must find businesses that can increase the commercial income by about 35 percent to pay for the repairs.

• CALTRANS is a large and capable organization with strong leadership and resources and clear policy direction to repair and rebuild damaged structures and to retrofit existing structures statewide. Three years after the earthquake, damaged freeway structures are still being demolished. Repairs take time, especially if there are unique or unusual structural and geotechnical conditions. Repairs and retrofit costing $100 million for the 1.6-mile double-deck I-280 viaduct linking the South of Market area to Highway 101 will not be completed until mid-1994. The $700 million I-880 project to replace the Cypress viaduct will not be complete until late 1997.

• In Oakland, 32 major buildings were damaged. Three years later, 16 still await repair or demolition. In some cases, owners can't pay for repairs or demolition, and the city is reluctant to pay, since the cost could exceed the value of the property. Oakland lost about 1,500 dwelling units for people with low incomes. Two years later, only 84 had been replaced.

Recovery is expensive for all parties. Federal disaster aid pays for only a small portion of the recovery, usually less than half. The various estimates of the damage and other losses from the Loma Prieta earthquake seem to be about $10 billion. The federal contribution through a variety of programs will amount to a bit over $2.5 billion, or about 25 percent of the total.

• The state of California imposed a 1/4-cent sales tax increase for 13 months to raise about $800 million. Insurance payments totaled about $1 billion. The remainder of the losses must be paid by local governments, businesses, and private individuals.

• Santa Cruz County recognized the need for additional funds and created a 1/4 percent increase in sales tax for six years. San Francisco and Oakland passed general obligation bond measures in part to finance repairs to damaged municipal buildings.

Recovery takes a long time, and it is expensive. Local government must hire the people and develop the skills to guide it through the process. Communities should anticipate the financial burden of disaster recovery and identify strategies to pay their share, even in lean times.

IF YOU CAN'T DO IT RIGHT, DON'T DO IT

The last three and one half years have demonstrated the difficulty inherent in implementing new knowledge and new programs. It is easy to say, "If we just applied what we now know, we could save money, buildings, and lives." But it is not easy to do.

Our professions are relatively knowledge-rich, but expertise-poor. A large segment of the professionals licensed to design our most significant buildings and infrastructure cannot competently apply the latest seismic design techniques. Yet these design and earth science professionals are responsible—and will be held accountable—for the vast majority of work that will either increase seismic risk if done wrong or decrease it if done right. Training pays off over time and can be accomplished without great cost.

In government, especially in California, it is difficult to hire and retain persons with seismic expertise. Multi-million dollar programs must be carried out using available staff, who generally lack specialized seismic expertise. Even though these people are intelligent, hard-working, and well-intended, they are invariably given significant responsibilities and tight schedules. They have precious little time and often no funds for professional training.

Contracting procedures, intended to provide qualified parties a fair chance at government contracts, make it difficult to distinguish among "licensed professionals." Laws now require that the cost of proposals and compliance with numerous quotas be given as much or more weight than demonstrated competence. Nontechnical managers who are responsible for personnel and contract decisions must rely on licensing laws for evidence of competence and on building codes for standards.

Expertise is critical to performance. The Loma Prieta earthquake caused much damage that could have been prevented. With more attention to detail, many of the monetary and function losses could be stemmed. Simple things could be applied, such as bracing shelves, ceiling systems, lights, and sprinklers. Owners need to know that if they don't insist on hiring people with demonstrated competence and if they don't insist on attention to detail and quality in construction, they are not likely to get it.

Code enforcement is critical to performance. The vast majority of our building stock is not "engineered." Because of the lack of professional expertise and the large number of buildings built without design professional involvement, we have a system that relies heavily on building codes and code enforcement to protect the public safety. Having building codes and enforcing them are two very different matters.

Public school buildings are one class of buildings that performed well, even those buildings built to older codes, which are now considered inadequate. Their performance showed that plan checking by experienced structural engineers and thorough, full-time inspection are more important than the codes themselves. Improving code enforcement by requiring rigorous plan checking, using expert review panels on significant buildings and structures and requiring thorough, "special" inspection, are proven techniques to reduce seismic risk and protect valuable capital investments.

You are the leaders of the professions responsible for seismic safety, but the overwhelming bulk of the work across the state and nation is done by others. These people are not here today. Most of our colleagues will never develop the needed expertise. Improving the knowledge of our practicing design and earth science professionals, code enforcement officials, and building contractors must be accorded the highest priority. To further reduce our risk, we need:

- to insist on professional training in seismic principles;
- to test for seismic knowledge in license exams;
- to consider seismic competence when selecting professionals and contractors; and
- to insist on meaningful code enforcement.

These recommendations can put our knowledge to work at relatively low cost, but they will go nowhere without your advocacy before boards of registration, in your professional societies, and before city councils and state legislatures.

THE ABILITY TO PREDICT DAMAGE

Before closing I want to continue with my theme of using knowledge to reduce risk through public advocacy by using one more example. In the months following the Loma Prieta earthquake, I listened to elected officials, news reporters, and other individuals criticize our statements that we could have foretold much of the damage. One member of the state Assembly stated during a hearing regarding funds to study the aftermath of the Loma Prieta earthquake: "You guys already know what's going to happen. You were right. Why do you want to study it? Just do something about it!" He voted "no."

After the Loma Prieta earthquake, I responded to questions from the press by saying it was common knowledge that the Marina District was prone to liquefaction; it was common knowledge that unreinforced masonry buildings were vulnerable to life-threatening damage; and, yes, it was common knowledge that older concrete structures were likely to lack the steel reinforcement and connection details needed to resist strong ground shaking. Virtually everyone in this room either did or could have answered in the same way. The press and the public were not impressed with our knowledge. They were shocked that we knew so much and did so little.

Because we already know or can readily find out where damage is likely, we must put this knowledge to work. Only by identifying vulnerable buildings and retrofitting them to appropriate standards can we prevent unnecessary losses and protect the buildings and functions we hold important. We must do a better job convincing policy makers to use our knowledge.

CONCLUSION

You are the elite from each of your professions. Your mere presence here tells me that you are motivated to learn and to apply what you know to your practice. Unfortunately, when Lloyd Cluff completes his closing remarks tomorrow afternoon, the knowledge gap between you and the tens of thousands of practitioners from all of our disciplines will be greater than ever. The gap between knowledge and public policy will be wider.

We need a new resolve to close these gaps in a systematic way. The successes and failures since Loma Prieta point to the absence of such a program, which is the foremost and most worrisome legacy of the earthquake.

Consider the points I have made:

• There is a growing gap between what is known and what is used and between the expert and the practitioner.

• Narrowing these gaps by enacting new policy depends on you; no one else is more knowledgeable or committed to the effort.

• NEHRP should be changed to emphasize implementation, incentives, and actions that reduce and manage risk.

• Disaster aid programs and our responses to damage need to accommodate irreversible changes and provide flexibility.

You and I must participate in advocacy. Each of us should ask whether our knowledge is being used. Ask how each new lesson can be used to lower seismic risk to life and the economy. Consider whether there is a policy mechanism to assure that the lessons will be applied by the majority of the professionals in your field. Consider the steps you can take to make a difference.

We all share a common need to improve earthquake risk reduction and management efforts. But the Loma Prieta earthquake and three and one half years of recovery have taught us that it will not be an easy task to accomplish this goal. We must change our strategy from sole emphasis on the development of knowledge, to one of integrating seismic risk concerns and knowledge into the mainstream activities of both government and business. We will succeed only when we contribute more knowledge to these activities and when we take advantage of the factors that motivate government and business. We will succeed only when seismic programs are no longer seen as separate programs and when you and I transfer and integrate our knowledge into mainstream private- and public-sector activities.

This is the legacy of the Loma Prieta earthquake. This is our charge.

2

The Geotechnical Aspects

G. Wayne Clough, James R. Martin, II, and Jean Lou Chameau

INTRODUCTION

Evidence obtained immediately following the Loma Prieta earthquake and in subsequent studies indicated a strong geotechnical influence on the observed behavior and damages. Much of the response could be termed "expected," but research following the earthquake has led to a refined understanding of previously defined problems and development of new areas of focus. For example, the earthquake allowed (1) a first-time "test" of sites that had been improved to resist liquefaction; (2) evaluation of soil density changes by comparing pre- and post-earthquake site test results; (3) direct measurement of effects of site amplification; (4) at least limited documentation of liquefaction-induced settlements and lateral movements; and (5) indirect measurement of the response of major landfills, underground structures, and reinforced earth retaining systems. Thus, considerable useful experience can be derived from the Loma Prieta earthquake for geotechnical engineering.

Although much has been learned from the earthquake, and more knowledge is to come, extrapolation of the information for the geotechnical community has to be tempered by the knowledge that special conditions ameliorated the damages. For example, even though up to 4,000 landslides occurred (Keefer, in press), such events were moderated by the effects of a four-year drought in Northern California. Other factors that should be considered in attempting to extrapolate lessons from the earthquake include the moderate size of the earthquake, the distance of the epicenter from large population centers and soils susceptable to liquefaction, the unique nature of the fault break and the relatively short duration of strong shaking, and the presence of low reservoir levels behind earth dams and embankments. These conditions make it essential that care is taken in using the lessons learned from the earthquake. If the duration of the event had been longer, water tables and reservoir levels higher, or the epicenter closer to San Francisco damages could have been greater, and what appeared to be successful performance could have translated into unsuccessful behavior.

The timing of this conference is well-matched to the discovery phase of the research on the Loma Prieta earthquake. It is notable that in the literature search for this paper many of the early, sometimes seemingly sensational, documents now seem dated. Their value for the future will not lie in the profundity of the insights developed but in the raw observations made of patterns of behavior. Clearly, the more recent publications, which reflect careful studies conducted in the intervening years since the event itself, are beginning to decipher properly the true causes of behavior. Also, with the publication of more research results, patterns are emerging that were not obvious before. This paper should be viewed as a summary of findings to date. Further useful results will undoubtedly be forthcoming.

OVERVIEW

It is useful to review some aspects of the Loma Prieta earthquake that are important to the geotechnical response associated with it. The M_s = 7.1 event (M_w = 6.9) was a moderate earthquake, with an epicenter located in the Santa Cruz Mountains, about 11 miles (18 km) from Santa Cruz and 60 miles (97 km) from the San Francisco Bay area. The causative fault rupture was bilateral, with a medial location of the epicenter. As a result, the strong shaking lasted only 8 to 15 seconds, shorter by as much as a factor of two relative to durations normally associated with an event of this magnitude. For the subsurface materials, this translates to a smaller number of cycles of loading than would have occurred otherwise.

The map in Figure 2-1 shows the position of the epicenter, major population centers, and locations of liquefaction-induced damage and landslides. As expected, there is a concentration of damages and landslides near the epicentral area, which reflects the high level of accelerations and steep terrain in this vicinity. South of the epicenter, in the vicinity of Santa Cruz, Watsonville, and Moss Landing, certain land masses were particularly susceptible to ground movement due to liquefaction and landsliding. Damages were also concentrated to the north of the epicenter in the San Francisco Bay area, where the type of earthquake motions and the soil conditions combined to produce site amplification and liquefaction in a heavily populated area. These effects are explored in more detail subsequently.

Table 2-1 lists peak horizontal accelerations and durations of strong shaking for a number of locations near the epicenter and in the San Francisco Bay area.

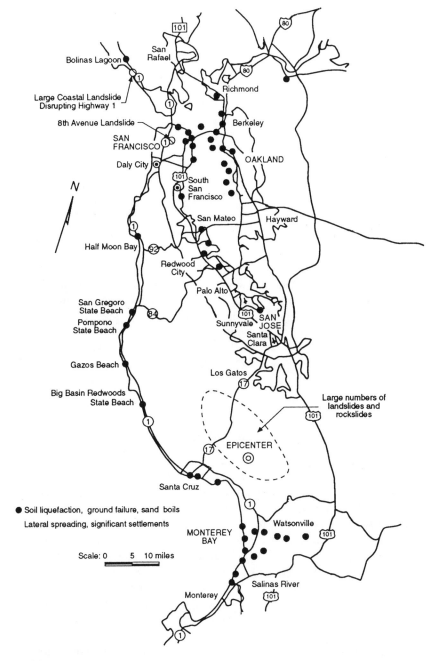

FIGURE 2-1 Regional map of earthquake damage due to liquefaction and landsliding (after Seed et al., 1991).

TABLE 2-1   Peak Horizontal Accelerations at Selected Sites in the Loma Prieta Earthquake

Location	Epicentral Distance (miles)*	Peak Ground Accelerations (gs)	Ground Condition
Epicenter	0	0.65	Rock
Santa Cruz	11	0.47	Rock
Watsonville	12	0.40	Rock
San Jose	14	0.25	Stiff Soil
San Francisco Airport	52	0.33	Fill/Soft Soil
Ricon Hill, San Francisco	63	0.10	Rock
Yerba Buena Island	64	0.07	Rock
Treasure Island	64	0.16	Fill/Soft Soil
Emeryville	65	0.24	Fill/Soft Soil

* (1 mile = 1.6 km)

The highest recorded accelerations were 0.6 g near the epicenter, and attenuation patterns of accelerations with distance from the epicentral region followed largely expected trends with some exceptions. Some 40 to 60 miles (64 to 97 km) from the epicentral region in the San Francisco Bay area, the peak accelerations varied from 0.05 g to 0.33 g, with the higher values associated with soft soil sites and the lower values recorded in rock and hard soil sites.

LIQUEFACTION

Occurrence and Recurrence

Liquefaction during the Loma Prieta earthquake was common near the shoreline of the San Francisco Bay and adjacent to rivers and bodies of water near the Pacific Ocean west of the epicentral region (Figure 2-1). There were few surprises as to the locations of liquefaction, since most of the areas where it was evidenced fit expected criteria for liquefaction. In a number of cases, liquefaction was accurately predicted prior to the earthquake (Clough and Chameau, 1983; Dupre and Tinsley, 1980). Recurrence of liquefaction in the same locations as in the 1906 San Francisco earthquake ($M_s = 8.3$) was not uncommon (Seed et al., 1991; O'Rourke et al., 1991). However, where damage patterns due to liquefaction in the earthquake mimicked those from the 1906 earthquake, the severity of liquefaction and damage from the earthquake was less than that associated with the 1906 event.

Well-documented liquefaction failures were associated with uncompacted, saturated, sandy fills in the central San Francisco Bay region (EERI, 1990). Figure 2-2 indicates where sandy fills are present on the eastern side of San

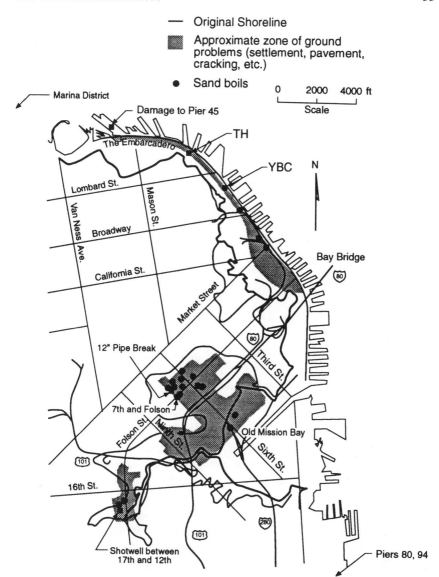

— Original Shoreline

■ Approximate zone of ground problems (settlement, pavement, cracking, etc.)

● Sand boils

FIGURE 2-2 Locations of waterfront fills in San Francisco and test sites TH and YBC. Liquefaction phenomena observed in Loma Prieta earthquake is noted (modified from Seed et al., 1991).

Francisco. The majority of these fills were placed in the late 1800s or the early to mid-1900s and consist of sands that were dumped or dredged into place and allowed to settle in suspension (Dow, 1973). Fills placed after 1950 tended to have been subjected to some form of compaction. Table 2-2 provides a description of many of the major fills and their placement processes. It is important to note that the fills that were dumped into the bay consisted of a wide range of materials, including rubble from construction and demolition. Another commonly dumped fill material was dune sand, a soil that was abundant in the early days of filling of the waterfront areas. Dune sand has a uniform, medium gradation and is largely free of fines. With time, sands were also dredged from sediments in San Francisco Bay. These materials typically contained fines, in contrast to the clean dune sands, and they attained lower fill densities than the dumped clean sands. During the Loma Prieta earthquake, differences in responses of fills created by different placement techniques was exhibited in areas like the Marina District (Bonilla, 1992; O'Rourke et al., 1990, 1991). The dredged fills exhibited a tendency to liquefy more readily than the dumped fills.

There were major fills around the bay that behaved well in the earthquake. In almost all cases, specific measures had been taken to compact the fills while they were being placed, or after placement. Those densified after placement are described in a subsequent section of this paper. The fills at the San Francisco Airport, and those at Foster City and Redwood Shores, were at least partially compacted during placement and exhibit medium to dense densities (EERI, 1990). In some cases, the fills also contain shells and are partially cemented. No significant liquefaction was found in these fills.

The Loma Prieta earthquake provided the first opportunity to assess the accuracy of regional liquefaction susceptibility maps (Tinsley and Dupre, in press). The susceptibility map in Figure 2-3 was developed by Dupre and Tinsley (1980) for the Monterey Bay region using the procedures of Youd and Perkins (1987). Liquefaction susceptibilities were based on the occurrence of a large event like the 1906 earthquake. In the Loma Prieta earthquake, the Dupre and Tinsley mapping accurately defined locations of major occurrences of liquefaction and lateral spreading. At the same time, broad regions within areas mapped as susceptible to liquefaction showed no response. This can be explained in terms of (1) the small size of the earthquake relative to the 1906 event, (2) lower water tables than those expected, and (3) local differences between grain sizes of deposits identified as liquefiable. The latter item was important in that flood plain deposits that were clay-rich did not fail, whereas areas of sand-rich tidal flat and abandoned channel deposits did fail. While broad mapping of liquefaction susceptible soils inherently has difficulty in capturing details, such as fines content or water table fluctuations, it is a valuable guide in identifying areas of potential problems.

TABLE 2-2 Behavior of Fill Soils in Central San Francisco Bay Region in the 1989 Loma Prieta Earthquake

Sites in City of San Francisco	Fill Type	Soil Type	Fill Density	Liquefaction Damage During Loma Prieta Earthquake*
Embarcadero	Unimproved End-Dumped	Fine Sand	Loose to Medium	Moderate to Minor
Hunter's Point Cofferdam	Unimproved Hydraulic	Fine, Silty Sand	V. Loose to Loose	Severe
Marina District	Unimproved Hydraulic,	Fine, Silty Sand	V. Loose to Loose	Severe
	Unimproved End-Dumped,	Fine Sand	Loose to Medium	Moderate
	Natural Ground	Find Sand	Medium to Dense	None
Mission District	Unimproved End-Dumped	Fine Sand, Rubble	Loose to Medium	Moderate to Minor
Pier 80, 84	Improved Dumped	Fine Sand	Medium to Dense	None
Pier 45	Unimproved End-Dumped	Fine Sand	V. Loose to Loose	Severe

Sites Outside San Francisco				
Alameda Island	Unimproved Hydraulic	Fine, Silty Sand	V. Loose to Loose	Severe
Bay Farm Island	Unimproved Hydraulic	Fine, Silty Sand	V. Loose to Loose	Severe

Sites outside City of San Francisco	Fill Type	Soil Type	Fill Density (V-Very)	Liquefaction Damage During Loma Prieta Earthquake*
Foster City	Roller-Compacted	Fine Sand, Some Cementation	Medium to Dense	None
Oakland Airport	Unimproved Hydraulic	Fine, Silty Sand	V. Loose to Loose	Severe
San Francisco Airport	Roller - Compacted	Fine Sand	Medium to Dense	None
Seventh Street Terminal	Unimproved Hydraulic	Fine, Silty Sand	V. Loose to Loose	Severe
Treasure Island	Unimproved Hydraulic,	Fine, Silty Sand	V. Loose to Loose	Severe
	Improved Hydraulic	Fine, Silty Sand	Medium to Dense	None

*NOTES: Minor = Slight lateral spreading and/or settlements, little surficial evidence.
Moderate = Minor lateral spreading and/or limited settlements, sand boils, etc.
Severe = Large lateral deformation and/or settlements, large & numerous sand boils, etc.
V. = Very

35

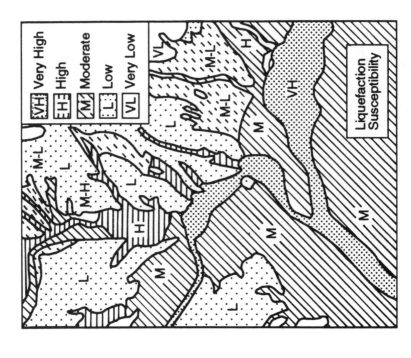

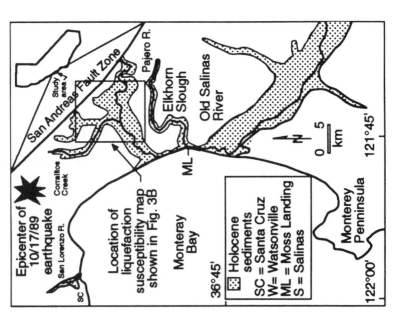

FIGURE 2-3 (a) Site map showing areas of Monterey Bay Region for which liquefaction susceptibility maps were developed; (b) Liquefaction susceptibility map developed for area indicated by inset in (a) (adapted from Tinsley and Dupre, 1992).

Settlements

Differential settlements due to liquefaction were widespread during the Loma Prieta earthquake. In level ground areas, the settlements were generally caused by a loss of fill volume during sand boiling and consolidation of the fills following pore pressure build-up. However, in instances where liquefaction occurred in the presence of a slope (even a very mild slope), downslope lateral movements in the soils caused settlements in the upper reaches of the soil mass that moved (see next section). Magnitudes of the settlements were essentially a function of the thickness of the liquefiable soils and the liquefaction potential of the soils. Liquefaction-induced settlements were believed to have caused failures in the San Francisco Municipal Water Supply System (Scawthorn et al., 1991); structural damages in the Marina District (Mahin, 1991), the South of Market area (Seed et al., 1991), Fisherman's Wharf and the Embarcadero (Chameau et al., 1991), and Treasure Island and the Oakland Port (EERI, 1990; Egan and Wang, 1991; Seed et al., 1991); and damages to highways and runways, for example, the Oakland Airport (EERI, 1990). It should be noted that even at sites where seawall and containment dikes were present, large settlements still occurred in hydraulic fills behind these support systems.

Accurate measured values of settlement due to liquefaction could not be derived from the information available to investigators. However, reasonable estimates could be made in some cases, and O'Rourke et al. (1991) were able to test existing methods of prediction of settlement caused by liquefaction. The procedures use results from Standard Penetration Tests (SPT) or Cone Penetration Tests (CPT) as input parameters. It was concluded that the methods worked well for clean sands but were not accurate in sands with fines unless corrections were applied to account for the effects of fines on the blow counts or cone penetration resistances.

Lateral Spreading

Lateral spreading was observed in most areas where significant liquefaction occurred during the Loma Prieta earthquake. In San Francisco, lateral movements were primarily associated with the sandy fills along the waterfront (Figure 2-2). These fills typically slope from 0.5 percent to 2 percent downhill toward the bay and are restrained by seawalls that run along the perimeter of the waterfront. Indications of lateral spreading in these fills during the earthquake were relatively minor. Much of the prominent pavement buckling in the central Marina District was attributed to oscillatory movements, not spreading. Except for an area near the marina where about 2 ft (0.6 m) of lateral movement occurred at St. Francis Spit (Taylor et al., 1992), permanent downslope (toward the bay) displacements of the fills were typically less than several inches. It must be kept in mind, however, that these movements were undoubtedly partially controlled by the presence of seawalls. Interestingly, Mitchell et al. (1991) have presented the thesis that the presence of large box culverts—typically 25 ft (7.6 m) wide and 30 ft (9.2 m) deep—buried along the perimeter of Marina Green helped to control the lateral spreading of the fills that liquefied in this area. This seems reasonable, and the stabilizing effect could have been amplified by the densification of the fills around the culverts that occurred as the sheet piles for the excavation support were vibrated into place (Clough and Chameau, 1980).

The relative lack of ground movements in the fills in the Old Mission Bay area was surprising, considering that lateral movements of up to 8 ft (2.4 m) occurred in this region during the 1906 earthquake (Youd and Hoose, 1978). The lack of significant lateral movements is partly attributed to the lesser magnitude of the Loma Prieta earthquake relative to the 1906 event, but other factors may have been at work. Possibilities include densification as a result of the 1906 earthquake, or that additional filling has occurred along the waterfront in this area since the 1906 earthquake. It is known that much of the rubble from buildings damaged during the 1906 event was pushed into the bay near the mouth of the old Mission Creek channel (Dow, 1973), and this could provide some buttressing support to the fills located farther inland.

Lateral spreading in the central San Francisco Bay region also occurred at Treasure Island, the Oakland Port, and other areas along the eastern bay shore. Similar to the occurrences in San Francisco, movements at these sites were also restrained by seawalls and containment dikes. Because the influence of the containment structures upon the observed movements is difficult to quantify, the actual deformation behavior of the soils is somewhat moot, and the movement data from these sites would be of limited use in the development of methods to predict lateral movements in liquefied soils.

Useful data on lateral spreading was obtained from the Monterey Bay area, where approximately 50 lateral spread sites were investigated by Tinsley and Dupre (1992). Lateral spreading throughout this region was strongly related to geologic facies. Approximately 95 percent of the spreads occurred in late Holocene fluvial point-bar deposits, fluvial channel deposits, and estuarine deposits. Beach and alluvial fan deposits rarely liquefied (see Figure 2-4). Geotechnical data from the field sites in the Monterey Bay area have yet to be sufficiently analyzed to develop relationships between the magnitude of the movements and the factors that controlled the movements. However, preliminary analyses suggest that the movements were not merely a function of simplified parameters such as slope angle, free-face height, etc. Although it was clear at all sites that the largest movements occurred near the free faces of laterally displaced slopes, there was no consistency in the "safe setback" distance from the free faces. At each site, the overall size of the soil mass that moved laterally was apparently controlled by the extent of the geologic unit in which the spread developed. Consistency for this selective behavior was confirmed by SPT and CPT measurements, which indicated soil conditions in the young fluvial and estuarine

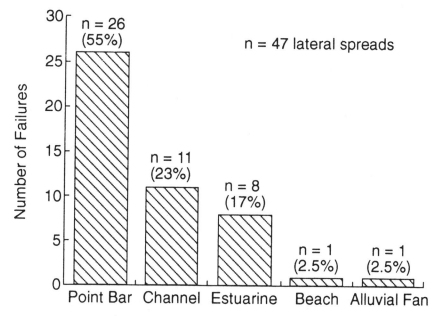

FIGURE 2-4 Histogram showing distribution of lateral-spread ground failures according to sedimentary facies for the Loma Prieta earthquake (after Tinsley and Dupre, 1992).

sediments to be the most favorable for liquefaction relative to older geologic units within the region.

Tinsley (1993) compared measured lateral movements in sandy soils in the Monterey Bay region with movement predictions obtained using the Liquefaction Severity Index (LSI) method of Youd and Perkins (1987). The results indicated the upper-bound estimates for lateral movements from the LSI method were too low. The LSI method predicts lateral movements based on earthquake magnitude and distance to the vertical projection of the fault rupture; it does not consider soil parameters. A more rigorous technique, developed by Bartlett and Youd (in press), as well as other prediction methods, have yet to be evaluated with the new field data.

Fill Densification As A Result of Earthquake Shaking

Following the Loma Prieta earthquake, SPT and CPT were conducted at several sites along San Francisco's waterfront, where similar tests had been performed in the late 1970s, which provided an opportunity to measure changes in density that were due to earthquake shaking (Clough and Chameau, 1983; Chameau et al., 1991). Two of the principal study areas, known as the TH and YBC sites, were located along the Embarcadero north of the Bay Bridge and Market

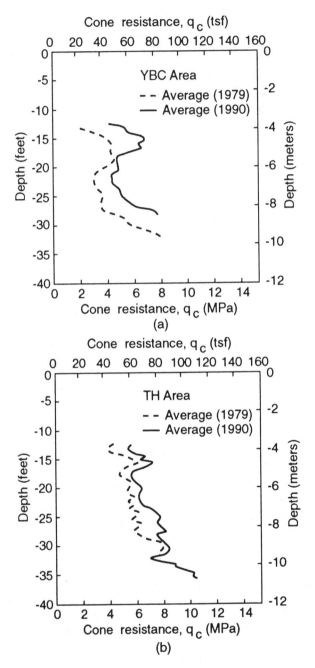

FIGURE 2-5   Average CPT tip resistances measured at the YBC and TH sites before and after the Loma Prieta earthquake (after Chameau et al., 1991).

Street (Figure 2-2). The soil conditions consist of dune sand fills about 30 ft (9.2 m) thick, which are underlain by soft recent bay mud; the ground water table is at a depth of 10 ft (3 m). Using conventional methods for liquefaction prediction, liquefaction would have been expected in the loosest portions of the fills at the YBC site but not in any of the fills at the TH site.

In Figure 2-5, the relative density profiles estimated for the pre-earthquake and post-earthquake conditions for the YBC and TH sites are plotted versus depth. Relative densities in the loosest sections of the fills at the YBC site were estimated to have increased from 45 percent to 60 percent due to the earthquake. Observations at the YBC site after the earthquake indicated minor sand boiling and lateral spreading and settlements. In contrast, no surficial evidence was found for liquefaction at the TH site, and no significant density changes were observed (average relative densities were estimated at 60 percent both before and after the earthquake). The implications from these findings are that moderate shaking and marginal liquefaction can be effective in densifying soils that are in a loose to very loose condition. Chameau et al. (1991) provide data for other sites to support the behavior observed along the Embarcadero.

Improved Ground

The Loma Prieta earthquake provided an opportunity to test the effectiveness of modern ground improvement techniques (e.g., vibraflotation, compaction piles, etc.) used to compact sandy fills and reduce their liquefaction susceptibility. There were no indications of liquefaction at sites where soils were improved using these procedures, although significant liquefaction damages occurred in adjacent areas of unimproved ground (Mitchell and Wentz, 1991). In spite of this positive behavior, it should be noted that the earthquake did not represent the design earthquake for most of the improved sites. Further, the ground motions at the improved sites were not significantly different from those at unimproved sites. Thus, while ground improvement may serve to inhibit liquefaction, it does not limit shaking of structures founded in the area.

SITE AMPLIFICATION AND RESPONSE SPECTRA

Site amplification is a term used to define the occurrence of ground motions at the surface of sites that are larger than those that would occur if the site were composed of bedrock. In the Loma Prieta earthquake, amplification occurred in the San Francisco Bay area at sites underlain by weak rock, stiff soil and soft soil (Borcherdt and Glassmoyer, 1992). The most dramatic effects were observed for sites with soft soil, and these are predominately found in the Bay Area around the fringes of San Francisco Bay. The soft soils were created by deposition over the past 10,000 years as San Francisco Bay was submerged under rising sea levels caused by the melting of the glaciers. These soils are termed locally as "recent bay mud," and they are generally near normally consolidated silty clays, although they can be either highly plastic clays or sands. The occurrence of soil amplification in the soft soils was not a surprise, since the technology available was able to predict this phenomenon. However, the experience of the earthquake provided insight into the degree of accuracy of predictive tools and important details about soil amplification.

The existence of site amplification can be seen in Figure 2-6, where ratios of peak transverse site accelerations to peak transverse accelerations for a nearby "bedrock" site are shown as a function of site conditions. All of the sites are located in the Bay Area some 40 to 60 miles (64 to 97 km) from the epicentral region. The harder rock sites show ratios of about one (as expected, no amplification), but the weak rock and soil sites have ratios above one, with average ratios of 1.2 for weak rock, 1.5 for stiff soil sites, and 2.5 for soft soil sites. The largest ratio for a soft soil site was 3.7 at the San Francisco Airport. In absolute terms, measured accelerations on rock sites were 0.1 g or less in the San Francisco/Oakland area. At instrumented sites that were underlain by soft bay mud, locations like Foster City, Treasure Island, or in the shoreline area of the East Bay, peak accelerations were 0.2 to 0.3 g. These accelerations were also rich in long period motions and, thus, possessed enhanced damage potential for weak structures (e.g., unreinforced masonry) and structures with long natural periods.

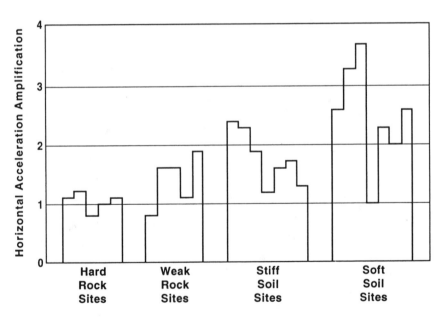

FIGURE 2-6   Acceleration amplification as a function of site conditions from Bay Area sites (adapted from Borcherdt and Glassmoyer, 1992).

Idriss (1990) was the first to report analyses of site motions that were conducted using conventional one-dimensional simulations based on piece-wise linear-elastic models of the ground response. He noted that this tool was successful in reasonably predicting response spectra of instrumental recordings at a number of soft soil sites if allowances were made for stiffer soil behavior and reduced damping in the soil model over earlier, accepted values. This conclusion was supported by Seed et al. (1991) and Dickenson et al. (1991), who performed analyses of a number of sites where observed and predicted response spectra were compared. Bardet et al. (1992) investigated the response of the soils in the Marina District of San Francisco using one-dimensional, two-dimensional, and three-dimensional analysis tools. The Marina District region is of interest in that the soils fill a basin with a three-dimensional shape where micro-tremor instrument recordings suggested an effect of this geometry (Boatwright et al., 1992). Conclusions from the Bardet et al. (1992) analyses were (1) one-dimensional analysis tools reasonably predict the response of the basin in the central regions, and (2) accounting for bedrock profile changes and wave propagation effects in two- and three-dimensional models can generate higher accelerations than are predicted in one-dimensional models. The results of all of the site motion analyses lead to several interpretations. First, it seems that for day-to-day design, the simplified one-dimensional models remain the most reasonable tool for site-specific analysis of ground motions, given the inherent unknowns about input motions and subsurface characteristics. However, for purposes of research studies of observed behavior, the more sophisticated analyses procedures would appear to warrant serious consideration.

While the site amplification effects in soft soils in the Loma Prieta earthquake could have been anticipated, the one-dimensional analyses showed that the largest amplified accelerations were not generated at the fundamental frequencies of the sites (Seed et al., 1991). Figure 2-7 shows response spectra determined from measured ground motions at Treasure Island (soft soil site) and the adjacent Yerba Buena Island (rock site). Both of the spectra exhibit peaks at periods of about 0.6 s (seconds) with motions for Treasure Island amplified about three times those of Yerba Buena Island. Seed et al. (1991) noted the amplification effect was generated by the focused energy in the input rock motion at a period of 0.6 s. With the predominant natural frequency of the Treasure Island site at approximately approximately 1.3 s, the peak spectral response that occurred at a period of 0.6 s was associated with approximately the period of the second mode of the site, not its first mode. At periods at or near 1.3 s, the response spectrum shows magnifications of four to five, but because the energy of the rock motions at this period was lower than that at 0.6 s, the magnitudes of accelerations at this period were not as large as those at a period of 0.6 s.

Figure 2-7 also presents the design spectrum proposed for the S_4 soil profile for purposes of comparison to the measured spectra. The S_4 condition was developed to provide design guidance for cases with deep soft soils. It is seen

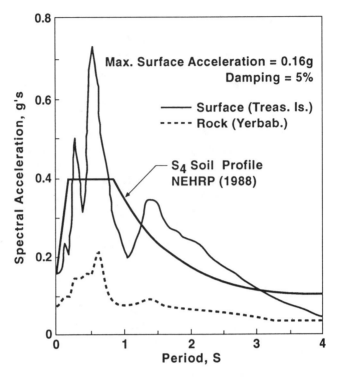

FIGURE 2-7   Measured response spectra at Yerba Buena Island (rock site) and Treasure Island (soft site) from the Loma Prieta earthquake with recommended S_4 design code spectrum (adapted from Dickinson et al., 1991).

that the measured response spectrum for the Treasure Island site significantly exceeds the design spectrum at periods where the maximum amplification occurred. Dickenson et al. (1991) show this effect applies for a range of spectra from soft soil sites in the Bay Area. This puts at issue the degree to which even the latest attempt to capture a conservative design approach for deep soft soil sites is adequate. As per the recommendations in such cases, a site-specific ground motion analysis is advisable.

SLOPES, FILLS, AND EMBANKMENTS

Landslides, downslope movements, and cracks developed in natural slopes, road fills, embankment dams, and landfills during the Loma Prieta earthquake (Figure 2-2). One fatality was attributed to a rockfall along the Pacific Coast, and damages were induced in over two hundred residences and fifteen earth dams. In addition, a major highway was blocked in two locations by landslide masses for extended periods of time.

Natural Slopes

A summary of the findings concerning the response of natural slopes in the Loma Prieta earthquake is to be provided in a forthcoming U.S. Geological Survey Professional Paper edited by Keefer (in press). Excerpts from this document provided considerable useful information for the authors. Keefer notes as many as 4,000 landslides may have occurred in the earthquake, with the most common types of landslides being small rock falls, rock slides, and disrupted soil slides. Deeper seated and more-coherent slumps and block slides were less common and more likely to be located near the epicentral region around Summit Ridge. The large number of landslides in the earthquake is not unusual, since it is estimated that about 10,000 landslides occurred in the 1906 San Francisco earthquake. The four-year drought at the time of the earthquake is thought to have prevented more-extensive movement and landsliding.

The natural slope landslides were concentrated in formations that had been previously identified as susceptible to sliding. Keefer reports that 85 percent of the landslides in the earthquake were located southwest of the fault rupture, mainly in the southern Santa Cruz Mountains, in the poorly indurated sedimentary rock formations. They were primarily composed of sandstones, siltstones, mudstones, and shales. Many of the landslides were in man-made cuts for highways, and two of these blocked different portions of California Highway 17 for several weeks. The highway cuts were materials that were highly fractured, and which were likely in a state of incipient failure prior to the earthquake.

Coherent slides were less common than the rockfall variety, but these were disruptive, since many were often located so as to impact residential areas or roads (Seed et al., 1991; Spittler and Harp, 1990). As many as 30 percent of the coherent slides occurred in road cuts, fills, or embankments, and most of these were small (Keefer, in press). The largest of the slumps occurred in the Summit Road area of the southern Santa Cruz Mountains. It involved up to 27×10^6 yd^3 of soil and rock with landslide depths of up to 300 ft (92 m). Trenches dug for purposes of identifying the slide planes for the Summit Road landslides showed that up to three displacement events in addition to that associated with the Loma Prieta earthquake had occurred over a period of 2,000 to 3,200 years (Nolan and Weber, in press). Cole (1991) attempted to use analytical procedures for slope and movement analysis to predict response of the coherent slides. Material parameters were determined from laboratory tests on small samples from the landslide areas. The calculations predicted movements that were smaller than those that actually occurred. It was recommended that more realistic predictions could be achieved using material parameters determined from back-analysis of previous slope failures. The effects of geometry and wave propagation should also be considered, since it was noted that block slides occurred in a number of locations at the end of prominent ridges.

In addition to the landslides, many ground cracks were opened in the Summit Road area. Some of the cracks were linked to adjacent landsliding, while others were indicative of incipient landsliding. Still other crack sets and fissures were found near the ends of prominent ridges. This may have been caused by wave reflection that would tend to occur at the face of a steep slope and by the known tendency for slopes to intensify motions near the face of the slope (Sitar and Clough, 1983).

A significant number of landslides and rockfalls were observed in the steep bluffs along the Pacific Coast (Griggs and Plant, 1993). While the coastal materials are often referred to as sedimentary rocks, they commonly are either highly fractured soft rock or cemented soils. Sitar and Clough (1983) showed the response of coastal bluffs varied depending on the tendency toward matching of the natural frequency of the slope and the characteristic period of the earthquake. For steep bluffs in cemented soils, the natural period for bluff heights of 25 m would fall in the range of 0.5 to 1, values that could coincide with those from the earthquake at distance from the epicenter.

Finally, while most of the landsliding occurred near the epicentral area, Seed et al. (1991) report a landslide that occurred in San Francisco in sands that damaged 30 homes, and one north of San Francisco on the west coast of the Marin Peninsula. These slides were located more than 60 miles (97 km) from the epicenter.

Engineered Embankments and Dams

Harder (1991) and Seed et al. (1991) reported on the behavior of earth dams that were impacted by the Loma Prieta earthquake, and much of the material that follows is drawn from these sources. There were 111 earth dams located within 120 miles (75 km) of the fault rupture of the earthquake (Figure 2-8). About half of the dams were built prior to 1950, and 21 of these were constructed before the 1906 earthquake. The height of the dams ranged from under 10 ft (3 m) to over 300 ft (92 m). Most were constructed of relatively homogeneous clayey soils, even those built before 1906. Five were constructed by hydraulic filling.

Of the dams that were built prior to 1906, none were damaged, a behavior consistent with experiences described for the 1906 San Francisco event (Seed et al., 1978). No significant damages were observed in any of the hydraulically filled dams, although Mill Creek Dam and Hawkins Dam experienced some minor cracking. Notably, the hydraulically filled dams were either located at a considerable distance from the epicenter or had dry reservoirs at the time of the earthquake.

Table 2-3 gives the key characteristics of ten dams that experienced noticeable ground movement or cracking in the earthquake. The majority of the dams exhibited cracking near the crest or in the materials adjacent to the abutment. Eight of the dams were earth/rockfill embankments and two were hydraulically filled. Not surprisingly, the closer the dam was to the epicenter, the more likely there was damage. Austrian Dam, which was only one mile (1.6 km) from the

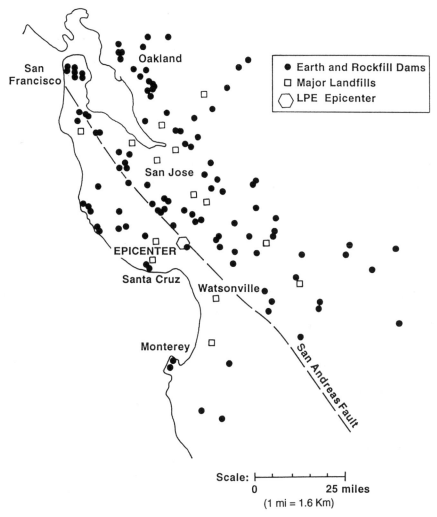

FIGURE 2-8   Location map for earth and rockfill dams and major landfills (adapted from Harder, 1991).

epicenter, sustained horizontal accelerations over 0.5 *g*, and it was the most severely damaged dam. Austrian Dam is 185 ft (56.7 m) high with 2.5:1 to 3:1 slopes, and it is composed of a compacted clayey, sandy gravel fill. The dam settled as much as 2.5 ft and displaced up to 0.5 ft upstream. The concrete spillway was extensively damaged, and cracking formed in the embankment longitudinally near the crest and transverse to the dam near the abutments. Water levels in five open standpipe piezometers set into the embankment materials

TABLE 2-3 Earth and Rockfill Dams Cracked in the Loma Prieta Earthquake (After Harder, 1991)

Dam	Type*	Date Completed	Height (ft)	Epicentral Distance (miles)**	Peak Ground Acceleration (g)	Comments
Austrian	Earth	1950	185	1	0.6	Moderate settlement, transverse and longitudinal cracking
Lexington	Earth	1953	205	2	0.5	Moderate transverse and longitudinal cracking, settlement
Guadeloupe	Earth	1935	142	6	0.4	Moderate cracking
Newell	Earth	1960	182	6	0.4	1–9-inch longitudinal cracks in U/S slope
Chesbro	Earth	1955	95	8	0.4	Moderate longitudinal cracking minor transverse cracking
Soda Lake	Earth	1978	35	10	0.3	Local slumping
Mill Creek	Hyd.	1889	76	12	0.3	Minor longitudinal cracking
Vessey	Earth	1945	20	13	0.3	Minor longitudinal cracking
Anderson	Earth	1931	72	21	0.2	Minor longitudinal cracking
Hawkins	Hyd.	1931	72	21	0.2	Minor longitudinal cracking

* Earth = earth/rockfill embarkment; Hyd. = hydrolically filled.
** (1 mile = 1.6 km)

rose after the earthquake, one by as much as 54 ft (16.5 m). It is not clear if these are representative of pore pressure increases in the embankment or other causes, but it was determined that several of the piezometer casings were bent in a manner indicative of spreading of the embankment near the lower portions of the fill. The damages to Austrian Dam are attributed by Harder to settlement and spreading of the fill. Had it not been for a low reservoir, Austrian Dam could have suffered greater damages.

Damages to most of the dams listed in Table 2-3 were caused by shaking and resultant settlement and lateral spreading of the embankment. However, the problems at Soda Creek Dam, and possibly Mill Creek Dam, were attributed to liquefaction. Soda Creek Dam retains tailings and was constructed with some saturated tailings left in the embankment section. It is believed that these materials liquefied during the earthquake, causing the embankment movements. Before the earthquake, it was thought the hydraulic fills within Mill Creek Dam were susceptible to liquefaction, since they were known to have low blow counts (the exact level of blow count could not be identified, since there was some controversy over the way in which the drill holes were supported). After the earthquake, small cracks were observed in the crest area of Mill Creek Dam, but no major damage occurred. The lack of damage was somewhat of a surprise, given the potential believed to exist for liquefaction.

Direct measurements of ground motions were made at eight of the dam sites, providing a resource for future studies. Harder reports that crest amplification effects appeared strongest in the cases where the base motions were the smallest. The reduced amplification in cases with higher accelerations was thought to be caused by increased damping or soil yielding in the embankment's soils that would occur if strains were relatively high. It could also relate to differences that exist between the natural period of the embankment and the characteristic period of the earthquake (higher motions are usually associated with higher frequencies). Two analytical studies of the measured embankment response concluded that finite element analyses could reasonably model the observed behavior (Makdisi et al., 1991; Sayed et al., 1991).

Sanitary and Hazardous Waste Landfills

A large number of major landfills and hazardous waste sites were shaken by the Loma Prieta earthquake. Table 2-4 presents a compilation of information on fifteen of these sites as derived from Sharma and Goyal (1991), Johnson et al. (1991), and Buranek and Prasad (1991) (see Figure 2-8 for locations). The landfills came in varied shapes, with some placed in canyons and others formed in mounds. The heights ranged up to 250 ft (76.7 m), with side slopes from 2:1 to 4:1. The Ben Lomond and Santa Cruz landfills were the closest to the epicenter and were subjected to horizontal accelerations of around 0.4 g. No major damages were observed at any of the landfills, although cracking was found at

TABLE 2-4 Major Landfills Affected by the Loma Prieta Earthquake (Adopted from Johnson et al., 1991, and Buranek and Prasad, 1991)

Landfill	Location	Type	Landfill Slopes	Effects of Quake	Peak Ground Accelerations (g)
Kirby Canyon Landfill	San Jose	Canyon fill	200 to 250 ft high, 2:1	No damage to slopes	0.3
Buena Vista Landfill	Watsonville	Modified gravel pit	up to 100 ft high, 3:1	Minor cracks or failures on trash slopes	0.4
Ben Lomond Landfill	Ben Lomond	Side hill fill	up to 150 ft high, 3:1	Minor cracking at contact with natural and on slope benches	0.5
Durham Road	Fremont	Mound	up to 90 ft high, 3:1	No damage	0.1
Newby Island	San Jose	Mound	100 ft, 3:1	No damage	0.1
John Smith	Hollister	Canyon fill	Gentle Slope	No damage	0.2
Ox Mountain	Half Moon Bay	Canyon fill	200 ft, 3:1	No damage to slopes; minor settlement	0.1
Dipauli/Vasco Road	Livermore	Canyon fill	150 ft, 3:1 to 4:1	No damage	0.05
Palo Alto City	Palo Alto	Cut and fill	60 ft, 3:1 to 4:1	Minor settlement at cut/fill contact	0.3
Santa Cruz City	Santa Cruz	Canyon fill	150 ft, 2:1	No failures of slopes; minor cracking at contact between fill and natural	0.5
Guadeloupe Landfill	San Jose	Canyon fill	up to 100 ft, 2.5:1	Minor cracking and downslope movement	0.4
Pacheco Pass Landfill	Pacheco Pass	Rock Site	125 ft, 3.6:1	Minor cracking	0.2
Marina Landfill	Marina	Area fill	up to 90 ft, 3:1	No damage	0.1
Zanker Road Landfill	Freemont	Area fill	up to 75 ft, 3.2:1	No damage	0.2

the contact between the landfill materials and the natural ground at sites closest to the epicenter. In computer simulations, Sharma and Goyal found that for landfills up to 50 ft (15.3 m) the base accelerations were increased at the top of the fill, but for landfill heights above this, the landfill acts as a "damper" and leads to accelerations less than those input at the base. Johnson et al. (1991) observed that the low density of the landfill material causes it to act as an energy absorber at the boundaries between the landfill and the natural ground. They also suggested that the landfill would offer considerable damping and that random materials within the landfill, such as boards and metal parts, would act as reinforcment for the landfill slopes.

WATERFRONT CONTAINMENT STRUCTURES, PIERS, AND LATERAL RETAINING STRUCTURES

The Loma Prieta earthquake provided the opportunity to evaluate the performance of waterfront containment structures including seawalls, rockfill dikes, levees, cellular cofferdams, piers and wharves, and lateral retaining structures. Rockfill seawalls constructed in the 1800s and early 1900s along the perimeter of San Francisco's waterfront and founded on bay mud generally performed well (notably the fills in these areas apparently were only marginally liquefied). At many locations where fills were located behind the walls, cracks were observed in the streets running parallel to the seawalls, suggesting minor lateral spreading toward the bay. Seawalls along the waterfront of the Marina District are supported on piles, and these exhibited little or no movement. In an adjacent area at the St. Francis Yacht Club, cobblestone seawalls that are founded on spread footings on sand fill suffered considerable damage, with movements of up to 1 ft (0.3 m) toward the bay (Taylor et al., 1992).

At sites where the restrained fills were strongly liquefied, rockfill dikes were not as effective in restraining movement of the fills as those along the San Francisco Waterfront. At Treasure Island, widespread liquefaction in hydraulic fills caused lateral movements on the order of 1 ft (0.3 m) of many of the levees surrounding the island. Similar behavior was observed at rockfill containment dikes along the East Bay shoreline in Oakland and Alameda.

Areas of the waterfront at the Hunter's Point Naval Yard were constructed on cellular cofferdams filled in the 1940s by dumped sands. Large settlements were observed in these structures (Chameau et al., 1991), and distress was evident at the interlocks of the cells. Complete failure occurred in one of the cells that formed the end of the cofferdam. This was caused in part by liquefaction of the fill and deterioration of the cell due to corrosion.

Piers 80 and 94, located along the eastern waterfront of San Francisco, behaved well in the earthquake. These piers are active container ports and were formed first by below-water dredging a wedge-shaped excavation in the underlying bay muds and subsequently by filling the excavated area by barge-dumping sands. The pier structures were supported on piles driven into the sands. The pile driving was expected to densify the sand fills. Tests performed after the earthquake showed the fills to average 60 percent relative density, suggesting the pile driving had densified the fills. Notably, this level of density may not be adequate to prevent damage in an event larger than the Loma Prieta earthquake.

At Seventh Street Terminal in Oakland, lateral movement and failure of a reinforced concrete wharf occurred. The wharf was supported by several vertical piles along the inboard portion of the wharf and by two rows of battered piles along the waterfront edge. The battered piles were designed to restrain lateral movements of the wharf, but they suffered severe damage at their tops during the earthquake, causing significant damages to the wharf. Further insight into the failure of this system is given in the next section, "Foundations."

A variety of lateral retaining structures were subjected to the effects of the earthquake. These included temporary-braced and tied-back walls, crib walls, soil-nailed and reinforced earth walls, and conventional retaining walls. Relatively little attention was given to these systems, since they behaved well in the earthquake. There were no reported failures, although some of the crib walls that supported fills along highway cuts were distressed slightly as a result of settlement of the fill (EERI, 1990).

FOUNDATIONS

The Loma Prieta earthquake provided the opportunity to observe the performance of foundation systems under earthquake loading and delineate differences in their behavior, as well as to test newer generation foundation systems such as base-isolated supports. Although some trends were identified and lessons learned, the fact that the Loma Prieta earthquake was of moderate severity meant that, in many cases, performances of foundation systems did not indicate limit behavior.

There were many cases throughout the bay region where damages to buildings and other constructed facilities varied by foundation type. The most prominent contrast was the behavior of structures founded on deep foundations compared with that of those on shallow footings. For instance, in the Marina and Mission districts of San Francisco, where liquefaction was prominent, structures founded on shallow footings were damaged, while those supported on piles embedded in non-liquefiable materials below the liquefied soils performed satisfactorily. A selective damage pattern that varied with foundation type was also observed in other liquefaction areas south of San Francisco, such as the Moss Landing area. Observations made during the earthquake also indicated that damages appear to be more likely if more than one type of foundation is used to support the same structure. For example, a building at the St. Francis Yacht Club in the Marina District was severely damaged due to large differential settlements between a newer pile-supported part of the structure and an older spread-footing supported portion of the structure. This was also demonstrated at Pier 45 where differential settlements of up to 1 ft (0.3 m) between the floors and walls of two large storage buildings forced closure of the structures. The floors of the building were founded directly on liquefiable sandy fill, but the walls and roofs of the sheds were supported by piles embedded into the bay mud underlying the liquefiable soils.

The earthquake also demonstrated the vulnerability of bridges founded on liquefiable alluvial soils to damage from lateral movements. Failures of this type were primarily located in the Monterey Bay area, near the epicenter. One of the most significant failures involved a seven-span bridge over the Pajaro River that had to be closed after the central pier of the bridge displaced laterally. The foundation soils consist of saturated, fine, and silty sands that liquefied during the earthquake and moved laterally, downslope toward the river. Evidence of liquefaction at the site was in the form of sand boils around the bridge piers, sand ejected from cracks running parallel to the stream, and minor slumping of the ground. Similar behavior was observed at the Neponset Bridge, a railroad bridge spanning the Salinas River. During the earthquake, liquefaction and subsequent lateral movement of liquefied soils toward the river bank caused several inches of lateral displacement of one of the bridge piers and a sharp curvature in the rail track. Interestingly, this bridge, which was built around 1903, suffered almost identical damages during the 1906 earthquake.

The failure of a bridge across Struve Slough on Highway 1 near Watsonville demonstrated that piles embedded in soft subsoils can lead to significant damages during earthquake shaking. The bridge was supported on concrete piles-columns that were embedded into very soft clay and peat underlying the site. During the earthquake, the piles settled and moved laterally, causing plastic deformation and shear failure of the top of the piles (EERI, 1990). There was also evidence of lateral movement of the underlying subsoils. Following the earthquake, gaps of 11.8 to 17.7 inches (30 to 45 cm) were measured between the piles and surrounding soils. The relative lateral displacement between the top of the columns and the base of the columns was estimated at about 30 cm. Although it was initially suspected that the foundation movements were related to liquefaction, there was little hard evidence of liquefaction at the site (no sand boils, etc.). Thus, it is uncertain what role, if any, liquefaction played in producing lateral movements at this site.

The earthquake provided insight into the ability of battered pile foundation systems to resist lateral loadings. As discussed earlier, failure of a concrete wharf occurred at the Seventh Street Terminal in Oakland due to the failure of two rows of battered piles installed along the outboard edge of the wharf to resist lateral loads. The piles were anchored into a dense sand layer, which underlies a rockfill containment dike and a layer of uncompacted hydraulic sand fill. Failure of the battered piles occurred when the containment dike and liquefied hydraulic fill material moved laterally toward the bay during the earthquake (EERI,

1990). Notably, none of the vertical piles at the wharf failed, nor did failure occur at adjacent wharves that were constructed with only vertical piles.

The earthquake provided the opportunity to test the performance of a base-isolated structure, the Sierra Point Bridge. The bridge, which is located on Highway 101 in south San Francisco, is 616 ft (188 m) long and supported on 30-inch (0.8-m) diameter concrete columns. In 1985, the bridge was retrofitted with seismic isolators that consisted of neoprene bearing pads that contained a central lead core. The isolators were placed at the top of the columns beneath the superstructure in an effort to reduce the seismic motions that would be induced in the structure during earthquake shaking. It was estimated that the increased period of the structure coupled with the hysteretic damping provided by the isolators would reduce the required seismic shear force by a factor of six (Mitchell et al., 1991). The bridge was designed for a peak acceleration of about 0.6 g at the ground surface. During the earthquake, a peak acceleration of about 0.09 g occurred at the base of the structure, an acceleration of 0.42 g was measured at the top of the columns below the isolator, and a reduced acceleration of 0.33 g occurred in the superstructure above the isolator. No structural damage was observed following the earthquake. Although the isolation system did reduce motions between the tops of the columns and the superstructure, it appears that the bridge did not behave entirely in the manner intended for a base-isolated structure. This is attributed to the fact that the abutments were not modified with a proper seismic gap to allow free movement of the structure on the isolator pads and the fact that the neoprene bearings were very stiff at the relatively low levels of strain induced by the earthquake.

One interesting phenomenon observed in the earthquake involves the relative motions of the supports of some extended structures (Earthquake Engineering Research Center, 1992). It suggests that even where soil conditions are similar, earthquake ground motions can vary between the supports of long-span structures, such as multiply supported bridges. The differential support motions presumably induce forces that are larger than those induced during uniform support motions. This effect is suspected to have been the primary cause of several bridge failures during the Loma Prieta earthquake, although additional research is needed to fully understand this phenomenon.

TUNNELS AND UNDERGROUND STRUCTURES

Underground infrastructure is reasonably well developed in the Bay Area. Most of it is related to lifelines, a topic that is the subject of another paper at this conference. However, it is appropriate in this work to consider the major underground structures such as those associated with the Bay Area Rapid Transit (BART) and the Clean Water Project (CWP) in San Francisco. The BART system is reasonably well known, and it travels underground in portions of the East Bay and San Francisco. BART passes beneath the bay in a submerged tube tunnel system that was placed in a dredged cut and subsequently filled to provide a cover over the tubes. The CWP is less well known. It consists of a massive network of near-surface culverts and tunnels that are linked to several pumping stations. The purpose of the CWP is to allow storage of storm runoff followed by pumping of the runoff to treatment facilities as water flows subside. Figure 2-9 provides a plan view of the location of the BART and CWP systems.

The designers of BART were aware of the need to accommodate earthquake effects in the underground structures (Kuesel, 1968; Douglas and Warshaw, 1971). The BART tubes pass through different types of materials, ranging from rock in some locations to soft bay mud under the bay and as the tubes enter land on the San Francisco side. Considerations were given to special earth loadings and earthquake wave-induced structural deformations for the underground sta-

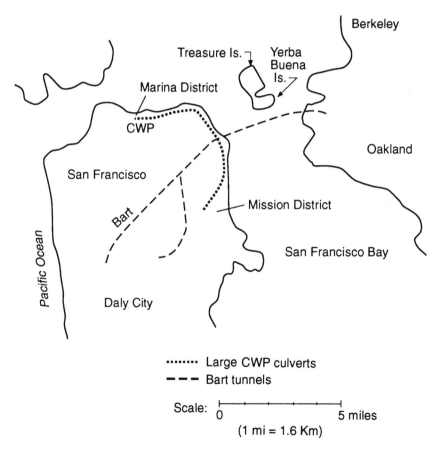

FIGURE 2-9   Locations of BART tunnels and large CWP culverts.

tions and tunnels. However, none of these resulted in any significant design changes, since allowances were also made for larger tolerances for structural stresses during earthquake loadings. Perhaps the most unique feature added to the BART system for earthquake protection was the special "seismic joint" used where the tubes rise from the bay crossing and tie into the land-based ventilation structures. Figure 2-10 shows a section through the seismic joint that allowed for up to 6 inches (15 cm) of relative longitudinal, shearing, and torsional movements between the tunnels and the ventilation structures. Teflon coatings were used on the contiguous steel surfaces to prevent binding. On the San Francisco side, the tubes lie in soft bay mud as they reach the ventilation structure, but on the Oakland side, the tubes pass from bay mud into natural sand at the ventilation structure.

The CWP culverts are large, typically 25 ft (7.7 m) wide and 30 ft (9.2 m) deep, and they lie at shallow depths (average depth of 5 to 10 ft [1.97 to 3.94 m]). The alignment of the culvert takes it along the Marina Green in the Marina District and along the waterfront from the Financial District to the Mission District. This places the culvert fully in fills that are known to contain loose sands and to be subject to liquefaction. The culverts were not pile founded, and they are subject to movement if large portions of the surrounding soils were to liquefy. Consideration of the likelihood of liquefaction led to design of the culverts for the fills acting in a full fluid condition.

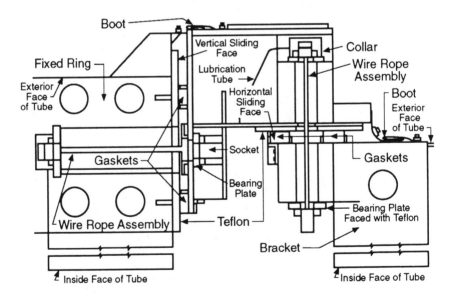

FIGURE 2-10   Schematic of seismic joint to connect BART tunnels to ventilation structures on San Francisco and Oakland sides of the San Francisco Bay (adapted from Douglas and Warshaw, 1971).

During the Loma Prieta earthquake, the BART and CWP were subjected to motions amplified by the bay mud, and sections of the fills in the Marina, Financial, and Mission districts at least partially liquefied around the CWP culvert. No damages were observed in either the BART or CWP systems after the earthquake, although there was evidence for small permanent movements in a few locations. Inspections of the seismic joints for BART showed no movement on the San Francisco side, but about 0.75 inches (1.9 cm) on the Oakland side (Chiu, 1993). The reasons for the Oakland side showing movement but not the San Francisco side are not obvious. Movements in the CWP culverts were observed at the joints between culvert boxes, and these could have been induced by the earthquake. On the other hand, it is also argued that these may have existed prior to the earthquake and were caused by general settlement. The lack of damage to the CWP culverts passing along the Embarcadero is to be contrasted with the severe damage that occurred to the Embarcadero Freeway structure that was above ground and founded on very deep piles. This illustrates the positive effects found in structures that are able to move with the ground. The CWP culverts as structures also possess a bending resistance that helps to resist local movements. It is not clear whether the CWP culverts would perform as well if extensive liquefaction were to occur under an event larger than the Loma Prieta earthquake.

LESSONS LEARNED FROM THE GEOTECHNICAL ASPECTS OF THE LOMA PRIETA EARTHQUAKE

There are many lessons that can be derived, and are being derived, from the experiences of the Loma Prieta earthquake. More is yet to be gained as further studies are conducted using the information that is still being obtained from investigations of the site conditions at locations where field behavior was documented. Care is required in extrapolation of the lessons learned, so that the factors that served to limit damages in the earthquake are properly included in the studies. This is likely not to be a problem with experienced earthquake engineers. However, there is a real danger that planners or managers who are unfamiliar with the technical details will underestimate the potential for damages in a larger event than the Loma Prieta earthquake or in different conditions with a wetter climatic condition. Consistent messages need to be delivered from studies of the effects of the earthquake. The key lessons learned from this event deal with liquefaction; site amplification; slopes, fills, and embankments; waterfront containment structures, piers, and retaining structures; foundations; underground structures; and areas for future concern.

Liquefaction

• Liquefaction susceptible conditions were accurately predicted using existing SPT- and CPT-based technology.

• Compacted or improved sand fills were made resistant to liquefaction in the Loma Prieta earthquake, but such improvements did little to reduce ground motions. The earthquake did not provide full design loading for the improved sites.

• There is a need for improved methods for predicting vertical and lateral movements caused by liquefaction, with allowances for the effects of fines on sand behavior.

• Broad mapping of liquefaction susceptibility identified areas of potential concern, but details of the subsoil conditions controlled the actual occurrence of liquefaction.

• Liquefaction, in a number of cases in the earthquake, represented a recurrence of liquefaction that occurred in the 1906 San Francisco event.

• Densification of sands after liquefaction occurred in a number of loose sandy fills. No densification was observed if the sands were of a medium density.

Site Amplification

• Although site amplification was most prominent for soft soils, it also occurred in stiff soils and weak rocks under specialized conditions.

• Site effects not only amplified accelerations but also changed the frequency content of the motions.

• Primary amplification can occur by matching of the characteristic rock motion with the second fundamental period of the site.

• Recent design spectra proposed for use with deep soft soil sites did not capture the maximum response periods in areas where site amplification occurred.

• One-dimensional site-response analysis tools were effective in explaining most ground motion issues, but more sophisticated methods were needed in cases where the underlying bedrock surface was non-uniform.

Slopes, Fills, and Embankments

• Slope failures left major transportation arteries closed for extended periods of time.

• Prominent topographical features (bluffs, promontories, etc.) apparently amplified accelerations and in some cases were subject to rapid failure because of resonance effects, wave reflection, and lack of confinement.

• Coherent, deep-seated sliding occurred only near the seismic energy source zone.

• Predictions of the amount of movement of coherent landslides under the earthquake loading based on laboratory-derived material parameters were not successful; back-analysis of existing slides appeared to offer a better alternative to obtain material parameters.

• Not all landslides induced by pre-Loma Prieta earthquakes were re-activated by the Loma Prieta earthquake, but incipient landslides were activated by the event.

• Well-designed earth dams sustained large levels of shaking, but localized damages could have led to problems if reservoir levels had been high.

• Valuable measurements were made of the acceleration response of earth dams that can be used in future studies.

• Sanitary landfills performed well in the earthquake, apparently in part due to their high damping capacity.

Waterfront Containment Structures, Piers, and Retaining Structures

• Seawalls limited ground movements in the presence of moderate liquefaction but were less effective where liquefaction was extensive.

• Liquefaction of fills contained within, or underlying, pier facilities was the primary source of damages to retaining systems and piers.

• Conventional retaining structures and more-recently developed systems, such as soil nailed walls, performed well, even where subjected to large accelerations.

Foundations

• In areas of prominent liquefaction, structures founded on shallow footings were damaged, while those supported on piles embedded in non-liquefiable materials performed satisfactorily. Notably, lateral movements of the soil surrounding the piles were typically less than one foot.

• Several bridges founded on alluvial soils in the Monterey Bay area suffered damage due to lateral movements of liquefied foundation soils.

• Structures that were constructed with batter piles to provide lateral restraint suffered heavier damage than adjacent structures with vertical piles.

• Seismic motions varied between the foundation supports of some long-span structures.

• Differential support motions presumably induce forces larger than uniform support motions and apparently caused substantial damage to several bridges.

Underground Structures

• Underground structures other than pipelines were not damaged by the earthquake, because they largely moved with the ground or provided bending resistance to potential lateral movements.

• The positive performance of the CWP culverts and the BART tunnels maintained the viability of essential portions of the infrastructure in the Bay Area. The lack of damage for the CWP culverts and the BART tunnels at the

San Francisco Waterfront was in sharp contrast to the damages to the above-ground Embarcadero elevated freeway system.

• The CWP culverts performed well even in the presence of liquefied fills, because they were designed for this loading condition. This has to be tempered by the fact that the Loma Prieta earthquake did not produce the large-scale liquefaction that might occur in a 1906-type event.

Areas for Future Concern

Areas for future concern include:

• the existence of liquefiable sandy fills in the waterfront areas of San Francisco Bay;

• lack of consideration of site amplification effects in design of many older structures;

• potential problems that continue to exist with landslides along the Pacific Coast and in the Coastal Mountain Range;

• the need for better understanding of how to incorporate site amplification effects into design spectra; and

• the need for re-assessment of what was perceived to be satisfactory performance in the Loma Prieta earthquake relative to what might occur in a larger event.

ACKNOWLEDGMENTS

The writers would like to gratefully acknowledge the assistance received in preparation of this paper. Information and advice was provided by T. L. Youd, D. Koutsoftas, M. M. Chiu, T. D. O'Rourke, M. S. Power, S. E. Dickenson, J. C. Tinsley, and D. K. Keefer. Assistance was also received from the Earthquake Engineering Research Institute, the Earthquake Engineering Research Center of U.C. Berkeley, and the National Center for Earthquake Engineering of the State University of New York at Buffalo. To all those who helped, we owe a significant vote of thanks.

REFERENCES

Bardet, J.P., M. Kapuskar, G.R. Martin, and J. Proubet. 1992. Site Response Analysis, The Loma Prieta California Earthquake of October 17, 1989—Marina District. U.S. Geological Survey Professional Paper, 1551-F. Washington, D.C.: United States Government Printing Office.

Bartlett, S.F., and T.L. Youd. In press. Empirical Prediction of Lateral Spread Displacement. Proceedings, 4th U.S.–Japan Workshop on Earthquake Resistant Design of Lifeline Facilities and Countermeasures for Soil Liquefaction," Honolulu, Hawaii, May, 1992.

Boatwright, J., L.C. Seekins, T.E. Furnal, H. Liu, and C.S. Mueller. 1992. Ground-Motion Amplification, The Loma Prieta, California, Earthquake of October 17, 1989—Marina District. U.S. Geological Survey Professional Paper, 1551-F. Washington, D.C.: United States Government Printing Office.

Bonilla, M.G. 1992. Geological and Historical Factors Affecting Earthquake Damage, The Loma Prieta, California, Earthquake of October 17, 1989—Marina District. U.S. Geological Survey Professional Paper, 1551-F. Washington, D.C.: United States Government Printing Office.

Borcherdt, R.D., and G. Glassmoyer. 1992. On the Characteristics of Local Geology and Their Influence on Ground Motions Generated by the Loma Prieta Earthquake in the San Francisco Bay Region, California. Bulletin of the Seismological Society of America. 82(2):603–641.

Buranek, D., and S. Prasad. 1991. Sanitary Landfill Performance During The Loma Prieta Earthquake. Pp. 1655–1660, in Proceedings, Second International Conference on Recent Advances in Geotechnical Earthquake Engineering and Soil Dynamics, Rolla, Missouri, Vol. II.

Chameau, J.L., G.W. Clough, F. Reyna, and J.D. Frost. 1991. Liquefaction Response of San Francisco Bayshore Fills. Bulletin of the Seismological Society of America, 81(5):2998–2018.

Chiu, M.M. 1993. Personal communication.

Clough, G.W., and J.L. Chameau. 1980. Measured Effects of Vibratory Sheet Pile Driving, Journal of the Geotechnical Engineering Division, ASCE, 106(GT10):108–1100.

Clough, G.W., and J.L. Chameau. 1983. Seismic Response of San Francisco Waterfront Fills. ASCE Journal of the Geotechnical Engineering Division, ASCE, 109( ):491–506.

Cole, W.F. 1991. Landslides Triggered by the Loma Prieta Earthquake, Implication for Zonation. Pp. 653–660 in Proceedings, Fourth International Conference on Seismic Zonation, Stanford, California, August, 1991.

Dickenson, S.E., R.B. Seed, J. Lysmer, and C.M. Mok. 1991. Response of Soft Soils During the 1989 Loma Prieta Earthquake and Implications for Seismic Design Criteria. Proceedings, Pacific Conference on Earthquake Engineering, Auckland, New Zealand, November, 1991.

Douglas, W.S., and R. Warshaw. 1971. Design of Seismic Joint for San Francisco Bay Tunnel. ASCE, Journal of the Structural Engineering Division, 7:1129–1141.

Dow, G.R. 1973. Bay Fill in San Francisco: A History of Change. San Francisco State University, M.A., Thesis.

Dupre, W.R., and J.H. Tinsley. 1980. Maps Showing Geology and Liquefaction Potential of Northern Monterey and Southern Santa Cruz Counties, California. U.S. Geological Survey, Miscellaneous Field Studies Map, MF-1199.

EERI. 1990. Loma Prieta Earthquake Reconnaissance Report, Earthquake Spectra, Supplement Vol. 6. Earthquake Engineering Research Center. 1992. News. Vol. 13, No. 2. June.

Egan, J.A., and Z.L. Wang. 1991. Liquefaction-Related Ground Deformation and Effects on Facilities at Treasure Island, San Francisco, During the 17 October 1989 Loma Prieta Earthquake. Pp. 57–76. Proceedings, 3rd Japan-U.S. Workshop on Earthquake Resistant Design of Lifeline Facilities and Countermeasures for Soil Liquefaction, Buffalo, NY, February, 1991.

Harder, L.F. 1991. Performance of Earth Dams During the Loma Prieta Earthquake. Pp. 1613–1629 in Proceedings, 2nd International Conference on Recent Advances in Geotechnical Earthquake Engineering and Soil Dynamics, Rolla, Missouri, Vol. II, March, 1991.

Idriss, I.M. 1990. Response of Soft Soil Sites During Earthquakes. Pp. 273–290 in Proceedings, H.B. Seed Memorial Symposium, Vol. 2.

Johnson, M.E., M. Lew, J. Lundy, and M.E. Ray. 1991. Investigation of Sanitary Landfill Slope Performance During Strong Motion From the Loma Prieta Earthquake of October 17, 1989. Pp. 1701–1708 in Proceedings, 2nd International Conference on Recent Advances in Geotechnical Earthquake Engineering and Soil Dynamics, Rolla, Missouri, Vol. II.

Keefer, D.K., ed. In press. The Loma Prieta, California, Earthquake of October 17, 1989—Landslides. NEHRP Report to Congress, U.S. Geological Survey, 1993.

Kuesel, T.R. 1968. Structural Design of the Bay Area Rapid Transit System. Civil Engineering, ASCE, April, pp. 1–6.

Mahin, S.A. 1991. The Loma Prieta Earthquake: Implications of Structural Damage LP03. Pp. 1587–1600 in Proceedings, 2nd International Conference on Recent Advances in Geotechnical Earthquake Engineering and Soil Dynamics, Vol. II, March, 11-15, Rolla, Missouri.

Makdisi, F.I., C.Y. Chang, Z.I. Wang, and C.M. Mok. 1991. Analysis of the Recorded Response of Lexington Dam During Various Levels of Ground Shaking. Seminar on Seismological and Engineering Implications of Recent Strong-Motion Data, Sacramento, Calif., May 30, pp. 10-1 to 10-10.

Mitchell, D., R. Tinawi, and R.G. Sexsmith. 1991. Performance of Bridges in the 1989 Loma Prieta Earthquake—Lessons for Canadian Designers. Canadian Journal of the Geotechnical Engineering Division, No. 18, pp. 711–734, January.

Mitchell, J.K., and F.J. Wentz. 1991. Performance of Improved Ground During the Loma Prieta Earthquake. Earthquake Engineering Research Center Report 91/12, University of California at Berkeley, October.

Mitchell, J.K., T. Masood, R.E. Kayen, and R.B. Seed. 1990. Soil Conditions and Earthquake Hazard Mitigation in the Marina District of San Francisco. Earthquake Engineering Research Center Report 90/08, University of California, Berkeley.

Nolan, J.M., and G.E. Weber. In press. Trenching Studies in the Summit Ridge Area. In Keefer, D.K., ed., The Loma Prieta, California Earthquake of October 17, 1989: Landslides and Stream Channel Change. U.S. Geological Survey.

O'Rourke, T.D., T.E. Gowdy, H.E. Stewart, and J.W. Pease. 1990. Lifeline Performance and Ground Deformations in the Marina During the 1989 Loma Prieta Earthquake. Pp. 129–146 in Proceedings, 3rd U.S.-Japan Workshop on Earthquake Resistant Design of Lifeline Facilities and Countermeasures for Soil Liquefaction, San Francisco.

O'Rourke, T.D., J.W. Pease, and H.E. Stewart. 1992. Lifeline Performance and Ground Deformation During the Earthquake. The Loma Prieta California, Earthquake of October 17, 1989—Marina District. U.S. Geological Survey Professional Paper 1551-F. Washington, D.C.: United States Government Printing Office.

Plant, N., and G.B. Griggs. 1990. Coastal Landslides Caused by the October 17, 1989 Earthquake, Santa Cruz County, California. California Geology, 43(4):75–84.

Sayed, H.S., A.M. Abdel-Ghaflar, and S.F. Masri. 1991. Parametric System Identification and Seismic Performance Evaluation of Earth Dams During the October 17, 1989, Loma Prieta Earthquake. Report No. CRECE-91-03, University of Southern California, Department of Civil Engineering, July.

Scawthorn, C.R., T.D. O'Rourke, M.M. Khater, and F. Blackburn. 1991. Loma Prieta Earthquake and The San Francisco AWSS: Analysis and Observed Performance. Pp. 527–540 in Proceedings, 3rd Japan-U.S. Workshop on Earthquake Resistant Design of Lifeline Facilities and Countermeasures for Soil Liquefaction, Buffalo, NY, February.

Seed, H.B., F.I. Makdisi, and P. DeAlba. 1978. Performance of Earth Dams During Earthquakes. Journal of the Geotechnical Engineering Division, ASCE, GT-7, July.

Seed, R.B., S.E. Dickenson, and I.M. Idriss. 1991. Principal Geotechnical Aspects of the 1989 Loma Prieta Earthquake. Soils and Foundations, Japanese Society of Soil Mechanics and Foundation Engineering, 31(1):1–26.

Sharma, H.D., and H.K. Goyal. 1991. Performance of a Hazardous Waste and Sanitary Landfill Subjected to Loma Prieta Earthquake. Pp. 1717–1725 in Proceedings, 2nd International Conference on Recent Advances in Geotechnical Earthquake Engineering and Soil Dynamics, Rolla, Missouri, Vol. II.

Sitar, N., and G.W. Clough. 1983. Seismic Response of Steep Slopes in Cemented Soils. Journal of the Geotechnical Engineering Division, ASCE, 109(2):210–227.

Spittler, T.E., and E.L. Harp, compilers. 1990. Preliminary Map of Landslide Features and Coseismic Fissures in the Summit Road Area of the Santa Cruz Mountains Triggered by the Loma Prieta Earthquake of October 17, 1989. U.S. Geological Survey Open-File Report 90-688, Scale 1:4,800.

Taylor, H.T., J.T. Cameron, S. Vahdani, and H. Yap. 1992. Behavior of the Seawalls and Shoreline During the Earthquake. The Loma Prieta California Earthquake of October 17, 1989—Marina District, U.S. Geological Survey Professional Paper, 1551-F, Washington, D.C.: United States Government Printing Office.

Tinsley, J.C. 1993. Personal communication.

Tinsley, J.C., and W.R. Dupre. In press. Liquefaction Hazard Mapping, Depositional Facies, and Lateral Spreading Ground Failure in the Monterey Bay Area, Central California. Proceedings, 4th Japan-U.S. Conference on Earthquake Resistant Design of Lifeline Facilities and Counter-measures for Soil Liquefaction, Honolulu, Hawaii, May, 1992.

Youd, T.L., and S.W. Hoose. 1978. Historic Ground Failures in Northern California Triggered by Earthquakes. U.S. Geological Survey Professional Paper 993.

Youd, T.L., and D.M. Perkins. 1987. Mapping of Liquefaction Severity Index. Journal of the Geotechnical Engineering Division, ASCE, 113:1374–1392.

DISCUSSANTS' COMMENTS: GEOTECHNICAL ISSUES

William Cotton, William Cotton & Associates

As a geologist, these are the three things I would like to expand upon from Dr. Clough's paper: (1) seismic zonation maps, (2) landslide reactivation, and (3) risk communication. Concerning seismic zonation maps, the ground did behave as expected throughout the Bay Area and throughout the epicentral region. There were maps in place before the earthquake that predicted ground behavior. The predictive kinds of seismic hazards, for which I have maps, include liquefaction-induced ground failures, landslide reactivation, and ground rupture in the Santa Cruz mountains and along the San Andreas fault. There is work being done by Roger Borcherdt at the U.S. Geological Survey and others about trying to predict the response of the ground to certain types of earthquake excitation. Amplification capability maps will come into greater use in the future. From these maps, the state of California has a new law in effect called the Seismic Mapping Hazard Act, which will take the three other hazards—shaking, liquefaction, and landslides—and try to produce maps based upon seismic response. Good ground behavior prediction maps were available for the Marina District, the San Francisco Bay margins, the city of Santa Cruz, and the Watsonville and the Summit Road region of the Santa Cruz mountains.

Concerning landslides, a large number of old deep-seated landslides in the epicentral region were reactivated during the earthquake. Most of these "coherent" landslides occurred in the Santa Cruz mountains. These slope failures provided a unique opportunity to evaluate the seismic slope stability and to test the dynamic slope stability methods currently being used in geotechnical practice. In addition, seismic displacement calculations were found to be very sensitive to selected yield coefficients, shear strength values, and acceleration-time history of the ground motion.

The most challenging risk, as I see it, is the transfer of geohazard information to individuals or groups that are charged with mitigating. Engineers and scientists have a poor record of packaging their research results and transferring their knowledge to the public. Thank you.

Maurice S. Power, Geomatrix Consultants

I would like to elaborate on two locations of liquefaction during the Loma Prieta earthquake. These were the Port of Oakland's 7th Street Marine Terminal and Treasure Island. Both are areas where hydraulic fill was placed through bay waters to create land.

During the earthquake the hydraulically placed sand fill liquefied at the marine terminal. As a result, the yard area settled by several inches, as did the rear crane rail, resulting in discontinuation of the crane operation. Lateral spreading movements caused extensive damage to the rear batter piles. The Seed-Idriss correlation for liquefaction potential predicted the occurrence of liquefaction. The Port of Oakland has reconstructed the facility replacing the batter piles with octagonal vertical piles. Of particular interest from the geotechnical viewpoint is that the ground has been improved. Post-earthquake ground improvement (vibro-replacement) was used to densify the sand to resist future lateral spreading. Comparison of pre- and post-improvement blow counts in the sand indicates that the sand was effectively densified at the Port of Oakland site using the vibro-replacement technique.

At the Treasure Island site, there was a general subsidence of about 4–6 inches (10.16-15.24 cm) during the earthquake, as well as spreading around the island perimeter. Again, the Seed-Idriss correlation agreed with the observations of liquefaction. Building sites on Treasure Island where pre-earthquake ground improvement had been implemented performed satisfactorily. There was evidence from pre- and post-earthquake survey measurements of an absence of lateral spreading movements at one location on the island perimeter where vibro-flotation had been performed, whereas adjacent unimproved areas experienced spreading.

Loma Prieta provided a wealth of recorded ground motion data. Improved attenuation relationships for estimating rock motions have been developed using these data. Correlations for assessing site response effects on ground motions have been developed. Knowledge of dynamic soil properties has been improved, and techniques for measuring dynamic properties have been evaluated. Analytical procedures for assessing site response have been found to give reasonable estimations of ground motions for the levels of excitation of the earthquake. Thank you.

Thomas Hanks, U.S. Geological Survey

The phrase "lessons learned" with respect to earthquakes first came into my consciousness 22 years ago, at the time of the 1971 San Fernando earthquake, when I was a graduate student at an epicentral distance of 62 miles (39 km). Back then, the National Academy of Sciences, which is sponsoring this meeting as well, put out a report entitled something like "Lessons Learned from the 1971 San Fernando Earthquake." We have been learning lessons from earthquakes ever since (and long before, of course), and this is appropriate for a scientific and engineering discipline that relies so heavily on observations, empiricism, and experience. And it is also true that a major earthquake in or near a major metropolitan area will be a learning experience for millions of people, most of whom don't know much about earthquakes.

Nevertheless, there is a certain déjà vu about many of the lessons of the Loma Prieta earthquake that are before us at this symposium. The effects of earthquake strong ground motion, for example, on unreinforced masonry, soft first stories, decayed timbers, bad foundations, hydraulic fill, and young bay mud hardly qualify as news here in San Francisco where these "lessons" had all been learned in 1906, if not before. These things keep happening, though, so we keep talking about them, but are we really making any progress in keeping these things from happening? And if not, why not?

With the latter question in mind, I learned a few things as a result of my own Loma Prieta experience. The first is that the American public, even in very affluent, very well-educated neighborhoods in the heart of earthquake country, is surprisingly uninformed about even the basics of earthquake occurrence, hazards, and risk. The second is that the practice of earthquake hazards reduction is very different from the theory of earthquake hazards reduction. In reality, earthquake hazards reduction is an intensely local happening involving large sums of money that hardly anyone wants to spend unless one absolutely has to, regardless of whether the source of funds is a government agency at the federal, state, or local level, or a neighborhood organization, or a private citizen. Third, an ounce of prevention in this business, like so many others, is worth a pound of cure, a largely unappreciated nicety, because it means spending money when you "don't have to."

I believe we should meet Tom Tobin's challenge: for all of us here and all that we represent, to take advocacy stands and more active roles to inform our citizen colleagues about what we know and to encourage our governing bodies to provide stronger incentives to practice earthquake hazards reduction in advance, so that potential hazards do not become real ones.

C. Thomas Statton, Woodward-Clyde Consultants

The Loma Prieta earthquake had some interesting timing aspects for my career in that we were developing some seismic design provisions for the New York City building code; the earthquake gave greater impetus to the team. But selling the idea of potential earthquake damage to a community that is essentially built, a community that has predominately poor ground on what land is left is not so easy. Our ability as group of scientists and engineers to convince the body politic that earthquakes occur and that earthquake engineering allows one to prevent a disaster is more difficult than perhaps on the West coast.

How does one translate the lessons learned from Loma Prieta to a different environment? One must transfer the context of the lesson as well as the craft itself. The context in the eastern United States and the western United States is quite different. For example, in the west, the sources of earthquakes are understood, but this is not so in the eastern United States. In the east, we deal with large segments of ground, and we call them seismic source zones. In the west, we deal with linear segments of ground, and we call them faults that produce earthquakes.

Another issue that needs to be addressed is that the seismic building code provisions are primarily derived from the California experience. So the rate of seismicity is important as we translate the lessons to the eastern United States context. When we look at earthquake design in the west for specific structures we find that the design earthquakes represent 80–90 percent of the maximum expected values coming from the maximum expected earthquakes. This is not true in the eastern United States, where the design values that are currently being looked at may represent 50 percent of the ground motions of the maximum expected event. So we may be designing for the same probability of occurrence of ground motion but nowhere near the same probability of being able to withstand the maximum event. In the eastern United States, the larger events that occur so infrequently may collapse structures, whereas in the western United States, the individual building performance will be significantly better.

In the east, the forward-looking view of lessons learned must be tempered by the fact that the east is largely built and built with buildings that have had no seismic attention paid to them, so the issue is how to apply retrofit lessons. Given the design event, we need to find a way to look backwards to a community that's predominately built. Thank you.

3

Buildings

Paul F. Fratessa

There were 63 deaths, $10 billion in direct and indirect losses and over 27,000 (California SSC, 1991a) structures damaged, and the knowledgeable consensus of experts is that there were few if any surprises. Those same experts caution against overconfidence, because the earthquake did not, in fact, provide a true test of earthquake resistant structures. Assuming these experts are correct, then results of a major earthquake close to a major urban area would be grim, and the Loma Prieta earthquake should be treated as a wake-up call.

Because there were few surprises, it seems logical that knowledge of building performance in earthquakes has advanced to the point where we can forecast building performance in the next event based on lessons learned not only from Loma Prieta but from all past earthquakes. If this level of knowledge is available, then it is clear that the mitigation of the hazard depends on the will and the means to get it done. This represents a challenge to the public to find the way to better safety. However, there is an equal challenge to the design and research community to be responsible to the public's trust and deliver the tools to make the mitigation effective.

The Loma Prieta earthquake also pointed out that the community was lucky and cannot be overconfident when viewing the prospects of a full service event. But there were indications of some potentially fatal flaws in the extension of the observed performance beyond the one event. Clearly the performance was good without surprises, but equally as clear is the perspective on what is not known and what the impact of that lack of knowledge might have on areas outside of California.

This paper explores the lessons learned from the Loma Prieta earthquake as they apply to buildings and then extrapolates those lessons to areas outside of California. The paper is based on lessons learned from both research and observation. These lessons were extracted from the research report summaries and from discussions with practicing professionals familiar with the Loma Prieta earthquake. The paper does not focus on the details of research or observations but on what are perceived to be the major lessons.

INTRODUCTION

Much has been written about the fact that there were "no real surprises" in the Loma Prieta event. Well-designed and well-constructed buildings performed well, while poorly designed and constructed structures did not. Implicit in this statement is the thought that someone can define "poorly designed and constructed." We can place unreinforced masonry buildings in this category, and they have long been identified as a problem. It is not as easy to identify other hazardous buildings as a class of structures. Some insight into what might be hazardous can be found by looking at damage statistics, which show that of the over 27,000 structures damaged, only 900 were reported to be of unreinforced masonry. The overwhelming majority of the damaged structures, classified by type, were wood-frame residential, a classification long ignored and presumed not to be potentially dangerous or the source of economic loss.

Functionality

When considering the true impact of the event, economic loss due to loss of functionality at the work place became a dominant focus. While the loss of building function did not surprise the engineering community, building owners and public agencies were surprised by the impact that loss of functionality had on communities. This was so much so that for the first time in California history, the governor acknowledged the need for functionality in state-owned buildings after an earthquake and recognized several key matters necessary to ensure seismic safety. In Executive Order D-86-90, Governor Deukmejian stated in part:

> . . . The Director of the Department of General Services shall prepare a detailed action plan to ensure that all facilities maintained or operated by the State are safe from significant failure in the event of an earthquake and that important structures are designed to maintain their function following an earthquake . . .

The order goes on to state that the plan should among other things:

> . . . (c) Seek independent review of structural and engineering plans and details for those projects which employ new or unique construction technologies; and
>
> (d) have independent inspections of construction to ensure compliance with plans and specifications . . .

For the first time, functionality was introduced as a performance goal, and quality control of design and construction were identified as keys to attain performance goals.

Acceptable Risk

Within this one document (EO-D-86-90) are the themes of some of the Loma Prieta earthquake's real lessons. Engineers must learn to recognize the need for post-earthquake functionality and the fact that our codes are intended as life safety provisions only. Legislation responding to the Loma Prieta earthquake required the California Seismic Safety Commission, in cooperation with the State Architect, to "develop a State policy on acceptable levels of earthquake risk for new and existing state-owned buildings and submit their policy to the Legislature for consideration by January 1, 1991." That document, Report No. SSC 91-1 (California SSC, 1991b) was submitted to the legislature in January of 1991 and has yet to be acted upon.

It needs to be understood that code concepts cover *normal* structures within the intent of the performance goals of the codes. Special considerations need to address special structures in order to obtain performance. Since keys to oversight of that special consideration are independent peer review and independent inspection, these should become mandatory for such projects.

Ground Motion

For a magnitude 7.0 event, strong ground shaking is usually expected to last about 20 seconds. The Loma Prieta earthquake was an unusual event as it had about 10 seconds of strong shaking and no clear surface-fault rupture. Seismologists are studying this phenomenon as they study the ground shaking that triggered numerous measuring devices yielding a plethora of valuable data. These data, when properly synthesized, will certainly yield valuable information toward predicting ground motion and damage distribution in future earthquakes. Certain anomalies are generally found in damage distribution and become the focus of intense study primarily by the earth sciences community. When these studies show that spectral design curves can be reliably modified from those currently in use, then this will be of distinct value to the design profession.

Despite the expressed lack of surprises, there was still much to be learned from the earthquake. While damage to types of structures and the effects of soft soils on building performance were predictable, there lingers some question as to whether longer and more-intense shaking generated by a similar magnitude earthquake closer to an urban area could have dramatically different results. Use of the ground-motion records and better understanding of the fault mechanisms should lead to better knowledge of earthquake risk outside of the Loma Prieta area.

Focus of Research

Disasters tend to generate a frenzy of activity in the research community. There is always a renewed interest in solving the problem and generally a limited amount of resources to fund the needed research. After the Loma Prieta earthquake, the August 1990 Earthquake Engineering Research Institute newsletter reported that the National Science Foundation and the U.S. Geological Survey awarded approximately $4.1 million in grants to do various studies. While learning from earthquakes, researchers should be studying the effects that are unique to that earthquake, yet it is interesting to note the focus of those grants. Twenty-four percent were focused on soil related topics, with an additional 10 percent on site-response issues. In sum, about 35 percent focused on geological and seismicity issues. Five and one half percent addressed evaluation and retrofit issues, and 2 percent addressed unreinforced masonry. Of note is that 2 percent of the awards money went to housing risk observations. The largest of the grants, 3.3 percent of the money, went to Risa Palm and Michael Hodgeson at the University of Colorado for work on "The purchase of Earthquake Insurance in California." Finally, 78 percent of the grants went to individuals associated with a university, and 10 percent went to practicing design professionals.

This summary of research awards is not a critique but is presented to focus the attention of the reader on evaluating what research seems to be needed as a result of the earthquake and where and how that effort might be best spent.

For research to be effective, it is essential that it be focused on the goals of a program. If one views research as the advancement of knowledge, then the studying of a subject for better understanding will in time be a benefit to the analysis and design of buildings. However, designers often have little patience with the research efforts, because the results do not translate directly into code language. There are short-term demands for research, such as the California Department of Transportation (CALTRANS) retrofit studies, which must be accomplished on a priority basis. Beyond those types of studies, there needs to be a more direct effort with a nationwide focus that will lead to advancements in knowledge in areas where it will have the most impact.

METHODOLOGY

This paper pursued two avenues of study in order to develop an overview of the lessons learned for buildings from the Loma Prieta earthquake. First, lists of the research papers available through October 1992 were reviewed. These lists were available through the National Information Service for Earthquake Engineering/Loma Prieta Clearinghouse Project. The lists were augmented by personal referrals to ongoing research, in some cases not directly related to the Loma Prieta earthquake. Second, a select group of practicing engineers from throughout the United States was polled for their thoughts on the significant lessons that can be learned from the earthquake.

The results of these two avenues of study (although not all inclusive) were then synthesized into broad categories and findings as they related to the earthquake. Viewing these findings, the appropriate lessons were then extracted as they apply throughout the United States. The paper is first organized into a discussion of the relationship of research to observation and codes and then into a series of findings, which emerged from the study. Within each finding, there are not so much lessons but discussion and opinions on the subject topic. The discussion draws on the study as well as the author's opinion. The paper concludes with the lessons extracted from the various topics that apply throughout the United States.

THE ROLE OF RESEARCH IN THE PRACTICAL LESSONS PROCESS

Numerous research programs have sprung up, often focusing on the same subject and each claiming to be that which will pull together and coordinate the activity. Along with those programs come the questions of how effective they are and how the research will be used. Then there is the latest buzz word in the technical world, "technology transfer." When someone has developed technical information containing knowledge on a subject, that knowledge is not automatically transferred by presenting a paper on the subject. Unfortunately, the mere availability of the knowledge does not necessarily transfer the knowledge or, in terms of seismic design, become useful in solving a design problem. Not all research is accepted out of hand, and it generally goes through a consensus process before it becomes an accepted theory. So how is research viewed by those who are in some ways the benefactors?

• *Researchers* seek analytical solutions or definitions of parameters from a postulated problem.

• *Engineers* seek a better definition of performance of materials.

• *Building officials* seek a prescriptive document, which is primarily black and white.

• *Building owners* wonder what they are paying for and expect a functioning building in the post earthquake environment.

• *Webster's Intermediate Dictionary* (1977) says:

    research . . . investigation or experimentation aimed at the discovery and interpretation of facts, revision of accepted theories or laws in the light of new facts, or practical application of such new or revised theories or laws.

As clear a definition as this may be, it is still heard differently depending on the listener. Researchers may view it as a way of life, while engineers complain that there is not enough information to provide precise answers. Funding agencies have a tendency to look at the complexity and sophistication of the research as an indication of its usefulness. And then, of course, the ultimate passing of the buck comes when information is not disseminated, or technology transferred.

In fact, research in the broadest sense is the backbone of our learning curve. Research does not always lead directly to an improvement in codes. Research, when viewed through experienced eyes, often improves the understanding of the earthquake phenomena by improved methods of analysis, improved understanding of capacities, and the confirmation of theorized performance characteristics. In looking back to the demand for research to mitigate the earthquake hazard, to be effective the research must have the potential to improve the understanding of performance but should not necessarily lead to a more complex answer to a simple problem. It should also be applied to the areas where the most impact can be gained.

Illustrations of the role of research generated by the Loma Prieta earthquake were the CALTRANS tests of retrofit confinement for concrete columns. Research was done for the need for short-term information for a limited number of special conditions (Roberts, 1990a, b; CALTRANS, 1991). Those tests have opened the research door to ways in which concrete can be confined in existing non-ductile frames. A consensus is not yet on the table for more wide use of that information. However, in this case, research was driven by a short-term, limited scope for a specific need. The extension of those findings has yet to come.

More broadly, research comes from the need to solve a yet unsolved problem or to expand our understanding of some aspect of seismic design. Engineers can often be impatient with the research community when researchers appear to be off on highly technical but broadly inapplicable subjects. Researchers must find a way to share their knowledge so that we are building a knowledge in a meaningful way.

There is no greater example of this than the potential windfall of information generated by the Strong Motion Instrumentation Program recording devices. A significant number of records are available for study through the California Division of Mines and Division of Geology. Not only do these records give researchers needed data for studies of the anomalies, but they give an opportunity to test results against current practice to see if it is really on the right track for a full-blown earthquake.

When viewing the broad potential of studying ground-motion records, there is a need for practical thinking. How should the spending of our research dollars be prioritized when considering what areas of research will provide the most return in the mitigation of the earthquake problem? As an example, much research attention is given to new construction and more advanced special designs of special structures; however, these structures are not the majority of the inventory in this country. These structures are also not the ones that generated the losses in the Loma Prieta earthquake. Should research follow the demand to mitigate? And if so, how does this theory relate to the uneconomical costs of retrofitting?

Is it appropriate for research funding to go to the specialty structural issues when they do not represent the vast majority of the structures subject to seismic loading? Perhaps it is time to rethink this issue. Should not the cost of providing the research for specialty construction fall on those who will most directly benefit from it, such as the owner? CALTRANS in essence has set that example. They are building high-use specialty structures and are doing the research to support their design concepts.

The Loma Prieta earthquake has left many opportunities for research. The emergence of research coordinating committees is a reflection of the concern for the need to coordinate efforts. As these efforts are coordinated with the intent to provide a forum for the dissemination of the results, it will be important to weigh the priorities of the research such that the results will truly be of maximum benefit in solving real problems and really mitigate the earthquake hazard. Patience on the part of practicing engineers will be rewarded if the coordinated research effort starts yielding usable results that, with the test of time, can improve the designer's understanding of the earthquake phenomena.

THE ROLE OF OBSERVATIONS IN THE PRACTICAL LESSONS PROCESS

The backbone of the California practice of seismic design is based on the observation of the effects of earthquakes on structures. These observations led to a concept of how buildings ought to be designed and eventually to the first code provisions. It is important to note that the observations needed to be tempered with research in order to bring about the advancement of rational design provisions. The ongoing process of observations tempered with research has continued to advance seismic codes. The downside of this process has been that as codes become more and more based on research, and as they become more complete, they also become more like a cookbook. This may be fine for those only concerned with the "does it meet code" syndrome; however, that approach does not relate back to the prime teacher of good performance. That teacher is not the code but observation.

Many people who had never experienced an earthquake or really seen the destruction of even a moderate event became observers during the Loma Prieta earthquake. Like the media, their attention was riveted on what failed or what looked spectacular. Observation of what did not happen can be as informative as viewing the obvious disasters. Much attention was focused on pounding between adjacent buildings. An example of that focus is the excellent work of Kasai and Maison (1990) in defining the parameters to assess the impact of pounding. The report discusses pounding damage and defines types of damage. It locates within a limited study area the various types and locations of observed pounding damage. What is not contained in the report is a true statistical analysis of the entire building stock within the study area and data on what pounding did not occur. It is also useful, and sometimes puzzling, to view separations between buildings that did not pound.

Observations tend to lead to quick code fixes in an effort to get rid of an obvious problem. The Loma Prieta earthquake showed numerous examples of soft-story structures. Soft-story structures are known to perform poorly. However, there were many soft-story structures that did not suffer damage or suffered damage in the soft story only, without damage to the upper stories. Is the appropriate response then to prohibit soft stories? Or should engineers evaluate the damping effect the softness has on eliminating force levels above? The codes, while not outlawing, have reacted by putting limits on soft stories that require additional analysis. In designing retrofits, many designers are ignoring the impact that stiffening the base has on the upper stories.

The Marina District of San Francisco was a media favorite during the earthquake. Observations were easy to make. There was liquefaction and there were soft stories. As noted above, a quick reaction to the soft stories was to outlaw them, however, the observation, astute as it may be, does not solve the problem of what to do with the existing soft-story structures that may have not been damaged. Was it the soft story or the liquefaction? Extending this line of thinking to liquefaction, we know we cannot outlaw it; however, what design provisions could be developed in areas of liquefiable soils? In fact, code provisions for liquefaction are not what is needed. Mitigating the impact of liquefaction on existing structures would be useful, but for undeveloped sites, liquefaction is a land-use planning matter much like the Alquist Priolo is for fault offset.

In summary, observations are excellent at focusing attention on critical issues. Often, short-term measures are warranted to mitigate the impact of what those observations discover. More often than not, the observations need some thought and discussion with the research community to see if the observations match past knowledge or if they are pointing at an anomaly. With time, the observations will lead to better knowledge and are often the trigger for research with a high potential for useful results.

PRINCIPAL FINDINGS

Seismic Risk

Sixty-three people are dead, and engineers are saying that structures performed about as expected. In excess of 27,000 buildings are damaged, and engineers are saying that is about what was expected. Public schools survived virtually unscathed, and again engineers say that is about what they expected. This was the first real confrontation with the performance intent of the modern seismic code. Before the Loma Prieta earthquake, "earthquake proof" was still a term that was used. Post-earthquake things have changed. The realization that buildings could be damaged and still be in complete compliance with the building code was now at hand. What were the expectations of the general public and the building owners?

Regardless of the expectations, there was a new reality. Buildings were not designed to be functional after earthquakes, they were designed only to be life safe. Building owners weighed the consequences of that, and soon the financial and insurance industries also saw the broad implications of such performance. It also became clear that retrofits for less than code levels were targeted for life safety only and that preservation of a historic resource was not the intent of applying the less than code approach. Retrofitting of historic buildings for less than *preservation performance* risks the loss of the assets.

Acceptable Risk Policy

The California Seismic Safety Commission was requested by a 1989 state law to develop a recommended policy on acceptable levels of risk in state-owned buildings (Fratessa and Turner, 1991). This law (SB 920) was signed by the governor about one month before the Loma Prieta earthquake. The commission responded with its recommended policy (California SSC, 1991b). This policy has had widespread attention, especially where it might apply as a general policy for all building owners.

The essence of the policy was to involve the building owner in the decision as to what level of seismic risk the owner wishes to accept. Clearly, nothing below life safety in new construction would be acceptable to the public. However, improved life safety for existing buildings was put forth as a possibility. A time frame of re-occupancy as a barometer of judging degrees of functionality was proposed. The bottom line was that engineers have traditionally assumed the role of the determiner of acceptable risk and it was now time for the owners to rightfully make those determinations. The engineer is responsible for the proper definition of those risks.

Owner Involvement

Clearly the Loma Prieta earthquake became the risk-enlightenment quake of the time. For the first time, owners were faced with the real decisions as to what level of risk they were willing to take and inevitably pay for if the cost was different. When this was extended to the post-earthquake repair and retrofitting of damaged buildings, it became not only the owners' issue but also that of the agency, financial interests, and the insurance company.

In first discussing acceptable risks, it became apparent that terminology was a key issue for understanding the subject. It seems that when the topic of seismic risk is raised, it is viewed from the perspective of the individual who often would place the individual's concerns or special interests at the top of the issue list.

To clarify the issue, the California Seismic Safety Commission defined the terminology. Later in this paper, this definition will become useful in discussing the impact of the earthquake, so it is developed briefly for reference.

Definition of Seismic Risk

Basically, the concept of risk is broken into two components: environmental risk and building performance risk. Environmental risk is that peculiar to the site, including the geology and seismicity. The type of shaking (spectral content), the intensity, the duration, and any special effects are considered. Liquefaction would also be an environmental risk, as would seismically induced landslides or soil instability. For any site, the seismological and geological communities must be able to define this risk.

Given the environmental risk, which is site specific, the performance risk has to be evaluated. This is based on the structure and involves the determination of the performance characteristics of that structure. The Loma Prieta earthquake moved that evaluation from just the safety or collapse mitigation to the determination of the level of damage and, therefore, post-earthquake occupancy potential. This latter functionality requirement puts a strain on the level of knowledge of the engineering and research community. Not only is the character of the ground motion critical but the performance characteristics of structural members in the inelastic range must be confidently known.

No single realization could have more significance on the design community not only in California but throughout the United States. As discussed later, this concept places the need and responsibility to make judgments on levels of acceptable risk on a clear platform.

Building Departments

Within California there is a considerable variation in capabilities of building departments. Depending on the size of the service area, there may be anywhere from a single, multipurpose individual who is expected to know and see everything to the large agencies with significant staff and, in some cases, even seismic-safety divisions. The degree to which the departments exert influence and control over projects also varies—often by choice but also by expertise, especially when the expertise is limited by the size of the department.

Tools and Technology Transfer

Departments are given tools to work with, which are called codes and technology transfer. In the specialized area of seismic design, it is particularly difficult to interpret, much less enforce, the code if one does not have a background in seismic codes. Literature is available and design seminars are offered in certain locations, but considering the other responsibilities of the job, it is virtually impossible for many officials to become completely conversant in seismic code matters. The ideal solution would be to have the code as clear as a cookbook and easy to interpret.

Building Damage Assessment

Immediately after the earthquake, various affected agencies responded to the crisis quite well considering the magnitude of the event and some of the pockets of localized damage. The state of California's Office of Emergency Services was called to test its well-developed response system and, as with any such event, was able to advance its knowledge. Given the competent initial response, the dust had to settle, after which new issues quickly arose. With the help of damage assessment volunteers and outside assistance from other agencies, the tagging of buildings for continued occupancy was accomplished. The tagging procedure and the lessons learned are discussed in Structural Engineers' Association of California (SEAC) (1991).

After this initial sorting out of the dangerous structures, a much more formidable task needed to be faced. That task was determining what to do with the structures that were tagged or structures that were in need of retrofitting regardless of the earthquake performance. In summary, the principal issues were

- What were the criteria for allowing the demolition of a building?
- What were the criteria for the repair of earthquake damage?
- When did the damage warrant repair, and when should it mandate retrofitting?
- What were the criteria to allow for the removal of red or yellow tags and therefore re-occupancy of the building?
- How does one deal with the sometimes conflicting requirements of the disaster assistance agencies, where the issue may be to rebuild the way it was?

There is a political issue surrounding the subject of tagging, but the California Office of Emergency Services demonstrated that there is a manner in which the actual tagging procedure can be effectively managed. Several outstanding papers are available on the experiences of emergency volunteers and their efforts to survey and tag potentially damaged buildings. The entire emergency response effort will not be reviewed here, although there are several practical spin-offs that were faced and can be learned from. Given the diversity of the assessment teams and the lack of knowledge of how they should be deployed, some local officials will be faced with a wide disparity in tagging consistency. Records show that a particular building in San Francisco was tagged four times by different individuals. Twice it was tagged yellow, then it was tagged red, and finally it was tagged yellow again before the tag was removed. In the interim, only some temporary shoring in one location had been placed. The Office of Emergency Services program is clear on training the volunteers and the authority under which their work is done. What happens after the initial response may depend on the municipality affected. The issue really is, "what is next after the tagging?"

The Initial Assessment

The initial pass in the damage assessment effort is to get hazardous buildings tagged so that they do not continue as a life-safety threat. The intent is to have more-detailed inspection where the structure is complex or the evaluation cannot be made in depth due to various constraints. The initial pass is most effectively made by teams trained to accomplish this task. This model is well established in California.

The second level of review is most effectively done by engineers experienced in seismic design and observation. With such experience, the evaluation should establish safety or use goals agreed upon in advance by the local jurisdiction. It is paramount that those who have had to deal with this before share their information with those who may someday face the issue.

A Question of Perspective

Clearly the Loma Prieta earthquake introduced to the engineering community a genuine anomaly when it was found that a building could be not a "good" seismic performance structure although it was not heavily damaged. Although a building may not comply with the modern code without significant damage, there is no criteria to mandate the retrofitting of the building. An engineer might be tempted to over-tag a building that the engineer feels "has no bracing," because the engineer knows the building is hazardous even though there was little damage. Those dealing with this matter should read the above referenced papers and have a workable plan of action and methodology in place prior to an earthquake.

Once the tagging program is in place, the criteria for the removal of the tag are as essential as the understanding of the consistent meaning of the tag to enforcement officials and the general public.

Ordinances Needed

Building departments were generally ill-prepared to deal with the post-earthquake repair or retrofit issues. Although some agencies within the state had in place ordinances dealing with these issues, this was not common, especially for the smaller communities in the Santa Cruz and Watsonville areas. The engineering community and the various state commissions had these issues on the list of work, but there was no consensus document available for each of these issues. As a result, emergency ordinances were hastily put in place to serve as a rational level of order. A prime example of this was the city of Oakland, which put in place emergency Ordinance 11217. This ordinance set forth the criteria to be used to determine whether a building was sufficiently damaged in the earthquake to warrant retrofitting as opposed to simple repair of the damage. If the damage was greater than a specified percentage, the building was required to be repaired in a manner to bring the building in compliance with the 1988 Uniform Building Code. If it was not damaged greater than that specified level, simple repairs were all that were required. The intent of the ordinance was to catch buildings that had major structural damage, and there was an intent to offer leniency for hardship as well as for structures falling reasonably close to the value of damage. The value set by the city was a loss of lateral capacity greater than 10 percent of its pre-earthquake capacity. The ordinance was supplemented by a procedures document (City of Oakland Memorandum, 1990), which set out the procedure for making the evaluation of both pre- and post-earthquake capacity.

Interestingly enough, the city has been witness to some of the most intriguing engineering manipulations by competent engineers for finding ways to make beneficial use of that procedure. However, like the manipulation of the building-period calculation in the more creative seismic design efforts, the procedure requires that one use the same logic before the earthquake as after the earthquake, thus resulting in consistent solutions.

Description of Need

Although Oakland's response is only part of the major puzzle, clearly some tools were found to be needed by all agencies. Some are technical issues while others are political issues involving the determination of acceptable risk for communities. Generally, they fall into the following categories:

- emergency standards and criteria for the demolition of a building for safety reasons;
- criteria for the determination of whether a building was damaged to a degree that it should be retrofitted rather than repaired;
- standards for the level of repair if repair only is mandated;
- standards for the repair for buildings to be retrofitted; and
- determination of the level of acceptable risk for the community involved.

In the immediate hours following a damaging earthquake, designated officials may be faced with a safety hazard from a building that is badly damaged and could collapse in aftershocks. Often, the older structures most likely to be in this situation may also be historic buildings. The options include clearing the hazardous area, shoring the building (with the possible danger to workers), or demolition. Most of this is political in nature; however, the potential danger for injury is part of the purview of the engineering community and is the subject of virtually no study or dissemination of knowledge.

The development of a consensus document for the evaluation of the level of damage that rationally should trigger the upgrading of a building is needed. The city of Oakland's program might be a good starting point. Because of the thought that return periods of significant events vary throughout the country, the procedure for evaluating the percentage of damage could be universally adopted while leaving the triggering percentage of damage to the assessment of risk to the responsible public agency.

Retrofit Standards

Given the triggering plateau or the simple need on the part of the owner to retrofit, the standards of retrofit need to be developed. Although standards are a first step, the techniques needed to implement the standards are also needed if the standards are to be effective. An example of this is unreinforced masonry. The standards adopted in the Uniform Building Code are now supported by commentary developed by the Structural Engineers' Association of California. The commentary assists the designer in determining what principles of detailing and analysis are anticipated in the code document. Still to be determined is the real effectiveness of these provisions. This will be learned only in an earthquake that shakes some of these retrofitted buildings. As in the past, observations will assist in modifying the document to get better or more-consistent performance. Holmes's study on unreinforced masonry structures in San Francisco is an excellent reference on the background of a city's issues on the subject (Holmes, 1990).

The broader field of other potentially hazardous buildings is not as far along as that of unreinforced masonry (URM) structures. This effort is underway by the Applied Technology Council (ATC, in press) and the state of California (California SSC, 1991c, 1992). The understanding of performance limitations of less than code-complying buildings is a difficult issue, since it involves the anomaly of the thought of having less than fully ductile elements or connections within a structure that might be subject to significant ground motions. A further issue is that nonstructural elements attached to those same structures may have the less than code level of detailing. A need to develop the tools for performance-oriented evaluations is in fact the demand. Unfortunately, the response to the issue seems to be an attempt to develop code standards as opposed to developing the analytical tools and the material properties that will give predictable performance values.

The development of the URM model code for bearing-wall buildings and the subsequent commentary was mandated by the need of building officials to have some standard to set forth as the minimum acceptable level of compliance. The user of the provisions is obligated to inform the building owner of the performance expectations implied by the URM code.

Evaluation of Risk: A Challenge

The Seismic Safety Commission's acceptable risk could clearly be used as the scaling factor for the political bodies in determining what levels of risk are acceptable in the future, again taking into account return periods. Remembering that the risk policy is based on building performance, including functionality, and that the engineering community is responsible for adequately predicting performance, it would appear that a major challenge in front of the research and engineering communities is the development of a methodology to predict levels of damage and extending that to imply levels of post-earthquake functionality. Of interest is the work of Freeman et al. (1993). This work is moving the design community past the strict compliance with code routine into real performance.

Once the engineering and research communities can develop the basis for the solutions to these issues, it will be necessary to determine the appropriate form in which to present this information to the building official so that the official can perform the fiduciary responsibility of protecting public safety.

Plan Checking and Site Inspections by Building Officials

Once the research has been developed and the industry brought up to speed, we are left with the dilemma of the building official. The official must enforce an ordinance that by its nature is inexact. Ideally, if all building designs were fully competent and flawlessly detailed the official would have no real problems. Follow that with detailed inspection by highly qualified inspectors fully familiar with seismic design as well as construction materials. That would make the official's job easier and might even diminish the need for anything as detailed as current seismic codes. This, however, is not the reality of the modern project, nor of the staffing capabilities of the average building department.

The building official, without even considering the Loma Prieta earthquake, faces a difficult problem with both the plan checking and site inspections of buildings. Certainly, budget plays a role in this. However, the public should demand that fees be charged that are adequate to provide proper levels of plan checking and site inspection. The Office of the State Architect (OSA) in California has a reputation for high performance for both plan checking and site inspection. Despite some complaints, the observed seismic performance record of structures reviewed and inspected by this agency is outstanding. Even so, it can be improved. In a legislative hearing about the Loma Prieta earthquake, surprise was expressed by a state senator when testimony indicated that plan check by any local building official was not the equivalent of that of the Structural Safety Division of the OSA.

Quality-Control Check Points

In all designs, there needs to be a series of check points to assure good quality control. The check points for design are the internal quality control of the engineer, the competence of the plan check, and the thoroughness and competence of the site inspection. Peer review is appropriate, especially when the design is complex.

Good construction implies that the design concepts go on the drawings so that a contractor can interpret them and then build in accordance with the drawings. The Loma Prieta earthquake uncovered situations where shear walls were left out of buildings and critical shear transfer elements were simply never installed. In addition, buildings were found to not comply with the seismic codes under which they were designed. So one can immediately blame the contractor or the designer, and yes, that is appropriate. But what about the last line of defense—the plan check for which each owner must pay a fee? For a moment, look at the situation for the building departments and the track records of success. In the Loma Prieta earthquake, post-Field Act school buildings had an outstanding record of performance. Anyone who has gone through the OSA process knows this is not an accident. Plan check with OSA is like having molars removed. One feels better about two weeks after the molar extraction. There is no doubt that OSA has gotten the job done. Yet in the California legislature, there have been recent efforts to eliminate the Field Act and OSA.

Some, but certainly not all, building departments in the state of California have plan check and inspection programs as well organized and effective as those of OSA. Given the present structuring of financial responsibility, it may be impossible, without new mandates, to improve laggard building departments. The lesson is that building departments, their inspectors, and plan checkers all have varying levels of capabilities. Given what will be called limitations, it is also appropriate to opine that there is no change from site to site in the need to get it right. As a result, there is a need to set some bottom lines for performance and find a way to help those agencies that need support when faced with quality-control issues involving plan checking and site inspection.

Some Action is Possible

The California Building Officials, the International Conference of Building Officials (ICBO), and SEAC should already be working together to determine how to implement a statewide program for quality control of plan checking and site inspection for seismic issues. The result would be of mutual benefit to each organization and to the public.

Strong Motion Instrumentation

The Strong Motion Instrumentation Program was the recipient of a major payoff in the Loma Prieta earthquake. The investment in instrumentation yielded numerous strong-motion records. The raw data developed from these recordings are available for study from the California Division of Mines and Geology (1990), and sets of recorded data have been utilized in the analysis of the performance of several buildings (McClure, 1991). Given this valuable information, it is interesting to note that virtually every engineer responding was concerned with the effective use of these data.

Knowing that the data were needed to advance the understanding of earthquakes and earthquake effects on buildings was enough to get the original Strong Motion Instrumentation Program well established. There is an equally tough challenge ahead to optimize the use of the data and research in a manner that will be most beneficial to the mitigation of the earthquake hazard. What might be the true benefit of research utilizing these data?

Purpose and Use of the Data

Clearly, if the data could be used to confirm or even postulate ground-motion characteristics in other regions, such characterizations could be used to assist in the overall evaluation of seismic risk. It would seem that an essential question must be addressed before the true value of the data will affect building design. That question is, what are the attributes that make a difference in structural performance, and are we now fully knowledgeable in the understanding of those attributes?

There seems to be a tendency to try to extract from the data a higher degree of accuracy than currently assumed for predicting the ground motion at sites affected by the ground shaking. The basic points of the effects of soft soils are understood in principal and addressed in code provisions. Small degrees of improvement in the accuracy of ground motion will not have a meaningful impact on the design process at its present state of knowledge except for on the most exotic designs. If the focus of the research is to identify conditions that are currently not recognized as critical issues in design, then the research should receive high priority. To the engineer who recognizes that this 7.0 event with 10 seconds or so of strong shaking was an anomaly, it will be interesting to see how the findings from this event will be extrapolated to other events not only in California but in other areas of the United States. What appears to be of major significance is the continued assessment of the relative, and to some degree, actual, risk of an event. Such information is vital to the total assessment of seismic risk to a community and would be useful in establishing priorities for funding as well as for mitigation programs.

Some Perspective

Regarding perspective on the interface between seismic design and the seismological experts, there does not seem to be any research being generated to identify the attributes mentioned above. As an example, do researchers understand what is more important in the successful design of a structure—the amplitude of the motion or the shape of the response spectra? Similarly, do earthquakes have predictable levels of energy input to various sites, and if so how would this be usable in the prediction of building performance? This bonanza of data should yield significant advancements in the mitigation of the earthquake hazard. Some thought regarding the use of the results could yield a planned program with tangible benefit.

Strong-motion records are being reviewed and analyzed to extract lessons on building response. Studies that correlate the data are useful for certain levels of ground motion, however, most studies are based on an elastic model and have difficulty accounting for inelastic or non-linear response. Again, the level and duration of the ground motion may make the lessons in response somewhat limited in many cases. Although there was a proposal at one time to fund a program for the utilization of the ground-motion data, the program has not yielded results. ATC 35 may fill that gap. There does not seem to be any systematic or organized program to extract the most important building-response data and apply them to useful lessons for seismic design.

Some work is appearing that uses the Strong Motion Instrumentation Program records to extract correlation, or lack thereof, from the current provisions with measured and observed performance. The work of Werner et al. (1992) is reported to offer some insight into what may be an anomaly relating to drift.

Building Performance

In general, buildings performed pretty well, in accordance with the designs used for the original construction. Of course, the media focused on the spectacular news to the point that many viewers on the East Coast thought that San Francisco was completely engulfed in fire and that Santa Cruz had been leveled. Yes, there was damage and death, and that cannot be ignored. However, there is cause to reflect on what did in fact perform well and focus on what was accomplished in California through the volunteer efforts of countless structural engineers when they, without funding, developed the first seismic codes. This same group, the SEAC, has continued that effort year to year basically without funding and has continued to perfect a workable and effective code. It is not perfect, but while reflecting on the Loma Prieta earthquake and what did not suffer damage, it is appropriate to understand that the level of seismic safety was achieved not by accident but by the unselfish efforts of the engineering community with the help of active researchers. Without that effort, the damage and loss of life would have been many times that experienced.

Still, there is a nagging view that despite the success, the earthquake did not truly test the designs, so one cannot extrapolate that we can relax now that we have experienced this event. Additionally, it is evident that although there were few surprises in performance, there was damage, and that implies that we have sufficient current knowledge to predict where the damage will occur in future earthquakes. If that is the case, we are faced with the traditional problem of having the will and the means to mitigate the hazard.

The Loma Prieta earthquake allows a number of observations about where the lessons of importance to future seismic mitigation may be.

Owners and the Public Expectations

In viewing the Loma Prieta earthquake where there were few real surprises, the clear lesson was that the owners and public were surprised, or at least given new insight, into what the seismic profession views as performance expectations. The public and owners had pre- earthquake expectations that unreinforced masonry buildings were hazardous, but they had the perception that earthquake-resistant design protected them from damage that might interrupt their businesses. There was also the perception that URM buildings would perform well if they had some limited level of retrofitting. The lesson learned was simple. The public is now aware that it takes more than minimum code compliance to have one's building operational after a significant event.

The first steps from the design community in response to this issue have been to address the need to be better at predicting damage and, concomitantly, the levels of response. Traditionally, the design focus had been life safety, which really just brought designs over the threshold to preventing collapse and tying a building together well. As noted above, that step saved countless lives, but a new demand has emerged and that is one of functionality. The Loma Prieta earthquake may have triggered that awareness, but it may well be misleading when judging damage potential from a magnitude 7.0 earthquake. Results would have been significantly different if the duration of shaking had lasted twice as long or if the quake had been centered in the northern section of the Hayward fault. Clearly, there is a mandate to develop better tools to predict performance as it relates to control of damage and functional requirements desired by the owner. Certainly, engineers designing structures in the post-Loma Prieta earthquake environment have a new obligation to inform the owners of projects as to the limitations of the minimum requirements of the code with regard to damage potential.

Performance, Design and the Code

In the Loma Prieta earthquake, well-designed, well-constructed buildings all performed very well. Evidence indicates that this performance is not an anomaly. Good design, good construction, and good quality-control do result in high levels of performance despite the inadequacies some may claim exist in our knowledge of earthquakes.

Breaking this down, it appears that for the Loma Prieta earthquake the requirements of the Uniform Building Code yield results consistent with its intent. With this in mind, the Loma Prieta earthquake was not necessarily a true test for all buildings.

As a result, there is a higher awareness of the true intent of the code on the part of the owners, and it is projected that in the future owners will ask designers to focus not on the minimum requirements of the code but on the need for post-earthquake functionality. If nothing else, this is music to the ears of the original seismic code writers who have been saying for years "remember these are minimums not maximums."

Safety Versus Post-Earthquake Usage

Tradition has basically led to the focus on the collapsed buildings. The Loma Prieta earthquake taught the public that they can be affected dramatically by the simple closure of a building, which takes away their home, place of worship, or place of work. Given this perspective, functionality has become a new issue. Since the codes are life-safety only, mere conformance to code no longer meets the demands of the public.

Focus of Research

A review of the types of research triggered by the Loma Prieta earthquake is interesting. The obvious attraction was the Bay Bridge and freeway structures. But this paper is on buildings. Again the focus was on the damaged structures that were generally unreinforced masonry or soft-story marina structures. The attention on masonry was obvious and direct, but it was known prior to the earthquake that these structures would not perform well. Being able to predict the poor performance should therefore direct the research attention to understanding how to repair or retrofit the buildings. Most attention has been focused on the URMs, however, it needs to be noted that wood-frame structures in the Marina District performed poorly. Some blamed this on liquefying soils. Whether it was the liquefaction or the soft story, or a combination of the two, little attention seems to have been paid to soft-story wood-frame structures, which were not damaged when subjected to similar ground motions.

Another perspective would be if one looks at the inventory of buildings suffering the most losses in the earthquake. In this case, it would clearly be wood-frame residential structures. This leads to the observation that to mitigate the maximum impact of the quake based on number of structures affected, one would focus on masonry and wood-frame structures, including residential structures.

Residential Construction

The residential problem was identified as a series of definable issues. The legislature mandated that the Seismic Safety Commission develop the *Homeowner's Guidebook to Earthquake Safety* (California SSC, 1992), which is basically a disclosure document that must be handed to the buyer of a pre-1960 house prior to sale. The document does a good job of identifying, in lay person's terms, the various hazards often found in homes. It is disappointing that the mitigation of those identifiable hazards is not mandatory prior to the sale.

The hazards to be mitigated for residential structures are relatively straight-forward. The consequences of not mitigating the simple hazards were pointed out in post-earthquake testimony. In testimony before the Seismic Safety Commission, a fire marshal was commenting on the need to brace water heaters and the potential for disaster if this is not accomplished. In the Marina District, gas fires were posing a serious threat. The fire marshal pointed out that it was lucky the gas problem was handled quickly before a fire storm situation developed and spread out of control. This was a dramatic testimony emphasizing the need for a simple mitigation measure.

Given that over 27,000 residential structures were damaged during the Loma Prieta earthquake, one would extract the logic that residential structures and the mitigation of the observed earthquake residential hazard would be high on the investment priority list of funding agencies. Two percent of the National Science Foundation (NSF) funding went to residential research, although the previously mentioned top award was on the topic of earthquake insurance in California. Only the California legislature took action in passing legislation to mandate the preparation and distribution of the homeowner's guide. It would seem obvious that the residential hazard is within reach of direct mitigation, and with some financial stimulus from governmental agencies and insurance interests, it is possible that this hazard could be wiped out.

The Profile of the Damaged Building

This brings the discussion to where the focus of research on buildings is, especially in the post-Loma Prieta earthquake time frame. The hazards are clear, and the majority of the hazards are identifiable as unreinforced masonry, non-ductile concrete, and residential construction anywhere in seismic zone 4. It should be noted that although the effects of soft soils were predictably more noticeable on soft soils for low-rise construction, there is no mandate based on observation to change the design approach to such buildings. Additionally, the profile of damaged buildings does not indicate that the problems were focused on new, tall, or exotic buildings even though this may not have been a significant event to test the true performance. It is also true that the majority of buildings constructed are not of that same category. The majority of the building stock in fact is low rise, and it was that stock that took a major hit during the Loma Prieta earthquake. If practicing professionals know where the problem lies, and generally agree that California's code provisions reasonably address the design of the majority of the problem buildings, then should not funding be likewise focused?

The Majority of Structures

Having said that, it is postulated that code-conforming buildings in the class being considered performed very well. If that is the case, then for the ordinary buildings without complexity, the proper application of the code may in fact provide a rational level of functionality. This leads to the conclusion that the functionality problems may result from buildings of such complexity or irregularity that they cannot be designed by code in the first place. Given that the code was intended for life safety only but for the majority of structures (the simple buildings) has yielded good functionality performance results, it may be inferred that all but the complex or irregular buildings are probably going to perform well even when conceding post-earthquake functionality issues.

For this *majority* building population the Loma Prieta earthquake did confirm that several factors are general predictors of performance. They are

- material age-deterioration;
- age of structure designed without codes;
- construction deviations; and
- improper application of the basic design concepts of earthquake resistant design.

Old Materials

Old materials put the issue of performance into the ballpark of how and if the materials can be saved while an effective system of lateral resistance is installed. The historical and social aspects of this debate are the subject of other papers, however, the engineering approach has been crystallized as a result of the Loma Prieta earthquake. As noted, there is a new awareness of functionality, not only on the part of owners but also in the insurance and banking industries.

To do the job right and bring a building up to expectations of performance can prove to be uneconomical. A response to this has been to allow some slack in strict code compliance and find a way, using all the materials, to make a building safe. Pre-Loma Prieta earthquake this was accepted, however, it is no longer professionally acceptable to fail to inform the owner of the realistic performance expectations of a design. Given this reality, owners are finding it difficult to retrofit at a functionality level, and without some financial incentives, the stock of truly historic buildings will be gone in the next major earthquake.

Unreinforced Masonry

Clearly, making a retrofit safe only for the occupants to evacuate now has a limited appeal, especially to bankers and insurers. Of the people killed by unreinforced masonry in the Loma Prieta event, all were killed outside of the building that caused the problem. No occupants were fatalities. This suggests that safety considerations go beyond the envelope of the building.

As expected, during the Loma Prieta earthquake, unreinforced masonry did not perform well. No research is needed to understand the basics of this. Faced with partially damaged structures, the methods of repair and retrofit become an issue. Two major areas of concern are apparent. First, what are the performance characteristics of the masonry, and what are the characteristics when the masonry is put into a framework? Second, when approaching retrofits, what will the performance be of the retrofit buildings, especially when the buildings have been stiffened and some of the high damping characteristics that, to a limited degree, have helped the performance have been removed?

The work by the SEAC Hazardous Buildings Committee in developing standards and commentary for unreinforced bearing-wall buildings has led to a broad introduction of the concept that the older materials, as frail as some are, do have some limited capacity for energy absorption. The performance of buildings utilizing such materials is not likely to meet the functional performance standards of a fully informed owner, however, as suggested by the Seismic Safety Commission policy on risk, improved safety may be the only economically viable alternative in some cases.

As noted elsewhere in this paper, observations of damage conclude that the damage to unreinforced masonry was as expected. Equally interesting and often unreported are the undamaged unreinforced structures. The supposedly successful performance of these buildings can often not be attributed to anything more than circumstances beyond the control of the structure, such as interaction with adjacent buildings or damping of a soft story.

The retrofitting of such structures, whether it be to repair damage, to meet safety requirements, or to meet performance expectations is not a simple code prescriptive matter. The understanding of the dynamics of seismic response and influence of inelastic action is essential to obtain a retrofitted structure with the desired performance characteristics. Equally as important is the attention to interstory drift at both the elastic and inelastic levels. Correlation with the recorded Loma Prieta earthquake results of actual building measurements should help with insight into this subject.

Non-Engineered Structures

A more difficult issue is facing the structures that have little or no engineered design and do not comply with the letter of the code. The code, or even our performance projections, are based on how engineers would like things to be, including idealized projections of material properties and building configuration. Unfortunately, existing pre-design buildings do not comply with those criteria, and engineers are forced to evaluate structures that have energy absorption characteristics that do not necessarily fit the mold. To address this, some new analytical concepts are being utilized that could offer significant insight into performance. These concepts focus not on capacity alone but on deformation and damping in the inelastic range.

Non-Ductile Concrete

No potentially hazardous category of building has had more attention focused on it then less than fully ductile concrete structures. Ban them? Use lower R values? Or as an alternative, perhaps understand their performance better? What risk is appropriate when fully considering the seismic risk of the locality? Studies by Gergeley have been aimed at projecting performance and specifically at revising the R values contained in code documents. Strictly from a performance point of view, it is essential that the capacity be understood and that the damage levels associated with these demands be developed.

The broader concern for non-ductile concrete, both existing and potential low-seismic-region design, has great broad impact. Clearly the Loma Prieta earthquake pointed to concern over those types of buildings. Building frames have been tested. These results, compared with the findings of the Loma Prieta earthquake ground-motion data, should give a better understanding of the effectiveness, or lack thereof, of non-ductile frames. Extending the ground-motion data into other parts of the United States will prove useful.

Research and Codes

It still remains that most of the applied research does not address the majority of the buildings but rather the exceptional buildings. However, only the knowledgeable should be the exceptional buildings, and the day to day fundamentals seem to already be well addressed by the basic codes. If that is true, then a lesson can be extended outside: the fundamental understanding of earthquake design is as important as any code. Codes have been getting more detailed and are covering issues that are not only difficult to understand but must be carefully applied in only the correct set of conditions.

Basic Principles

Engineers experienced in seismic design agree that adherence to basic principles such as the providing of complete load paths and the rational transfer of forces in connections is essential to good seismic design. The code should support those basic principles by providing material parameters to allow the accurate evaluation of the strength of the materials used. The limits on the use of those parameters are equally important. The user of the code should then be able to read the code, determine the material strengths as well as the limitations of use of the parameters. When the limitations become more complex, the reader may lose site of the basics and become absorbed in the complexity of the code provision.

Shear stress in concrete is an example. Historically, shear in concrete was a simple calculation that added the shear carried by the concrete to the shear carried by the reinforcing steel to give the total shear capacity of the section being analyzed. This was a simple straightforward computation and gave a basic sense of how loads were transferred in concrete. The modern code provisions for shear stress in concrete offer much more complex methods for evaluating shear capacity. Without study or research, the reader of the code may not fully understand the basics on which the provision is founded.

When viewing the shear wall damage, it is apparent that basic application of shear wall principles would have been a more effective mitigation measure than the application of more-complex code provisions. Stated another way: would the application of more-complex concrete shear wall provisions have improved the performance of buildings in the Loma Prieta earthquake? Would the nuances of the more complex provisions really have a significant effect on the performance when compared with the other performance parameters?

In short, codes may be getting too complex, may be trying to cover too much, and may be trying to cover all situations in a prescriptive manner. In principle, the code should address the largest inventory of buildings with basic provisions that cover the simplest of structures. As structures become more exotic, a higher level of design expertise is required. These more exotic structures should not be the focus of the codes but should require more sophisticated analysis, something the codes will never be able provide in a prescriptive manner. The review of such structures should be accomplished by independent peer review and not dumped on the building official.

Soils and Building Performance

A second lesson is clear. Loma Prieta was not the earthquake that typified the San Andreas fault, and it gave us a little wrinkle. It was a Richter magnitude 7.1 with a relatively short duration. However, Bruce Bolt has developed a model that correctly identifies the pockets of damage experienced during Loma Prieta earthquake (Lomax and Bolt, 1992). It remains to be seen if such a model applied with sources located at other sites along the San Andreas or other faults will lead to any insight into the need to make revisions to the current static-force procedure or whether it will assist in providing more-accurate site-specific information for those using more sophisticated analyses. Equally important would be the effectiveness of such a model for fault conditions outside of the California scenario.

While attention is being focused on site-specific data as it is traditionally defined, there are perhaps other issues in need of study. The earthquake gave us some newer terminology such as "focused energy" and "near field effects." In today's design approaches, there is little opportunity to address either of those notions. In fact, it is clear that stepping from such concepts to meaningful design parameters is a major issue when considering the effectiveness of research in the mitigation of hazards. While the parameters of sites are described in terms of peak accelerations and return periods, a designer must focus attention on the spectral shapes as much as the ordinate of the spectrum. The complete understanding of earthquake design must include the understanding of inelastic response and how the inelastic softening of a building affects performance and the control of damage. The future may hold yet another parameter; the total site energy related to a given event. There is need for the environmental side of the risk equation to equate with the building-performance side so that the data from the Loma Prieta earthquake is effectively utilized for building performance in a manner that will advance our state of knowledge and not just provide interesting information.

It would be unfortunate to have the data of this limited event extended to rationalize modifications in the basic approach to static design, which in fact demonstrated good results in the Loma Prieta earthquake.

CONCLUSIONS

Acceptable Risk

For any location within the United States, there are data that describe the environmental earthquake risk as best we know it today. The national effort to improve that data base was enhanced by the Loma Prieta earthquake. It is imperative that the National Earthquake Hazards Reduction Program (NEHRP) take a lead role in coordinating that effort such that there is a continuing improvement of that data and its application to seismic mitigation. The resultant knowledge must then be transferred not only to the design profession but also to the decision makers.

Such decision makers should then be looking to the design professional to define the building-performance risk associated with a given site. That performance must be reasonably predictable for the normal buildings, and may well require special studies for the complex or irregular buildings. It is opined that code compliance will not and never was intended to deliver that level of information. The Loma Prieta earthquake has alerted the owners, and they should now be demanding rational predictions of building performance based on post-earthquake functionality.

This rationale places a new burden on the research and engineering communities. However, in so doing, the earthquake has brought new awareness and, therefore, new responsibility to the developers or owners of buildings. This concept is that of acceptable risk, and it is a universal issue wherever earthquakes could affect communities. Some education of the public on this subject is mandated.

Fundamentals, Not More Code Provisions

When the consensus that there were no surprises was voiced, some may have wondered how much was known or maybe whether the code was right after all. Don't count on it. The message was that well tied together, rationally designed buildings worked fine, perhaps despite the code. Basic fundamentals of seismic design seem to prevail where good performance is found. Where such basics are missing, poor performance is evident. Clearly, older buildings constructed prior to the understanding of seismic performance generally lacked the basic elements of good seismic design.

The lesson in this case is the same lesson as learned before in other earthquakes. The simplest of codes could say: tie the building together, provide complete load paths, and design it to meet requirements. But codes have become more complex, and perhaps in the effort to cover all types of buildings, they have obscured the fundamental of the majority of the constructed buildings.

The lesson is that it is necessary to keep the basics simple and understandable. Professionals (doing the design of the majority of the structures) need education in the simple principals of seismic design. Perhaps research efforts, along with the code bodies, ought to consider solving the majority problems on simple and basic terms and leave the more complex issues to be addressed by those proposing to design such structures.

Peer Review for the Complex Projects

Considering the more complex structures, if the above approach were to be applied, perhaps most everyone would be happy except the building officials. If the more complex structures use concepts not specifically stated in the code, how are officials to enforce compliance? There were numerous structures damaged in the Loma Prieta earthquake, which for legal reasons will go unnamed, and in which there were design or construction errors that technically should have been found within the purview of the building-department program. This is not a criticism of the competence of the individuals but a statement of a problem with priorities that do not allow agencies to fund the necessary staff to accomplish a fundamental step in the construction process.

If the system of enforcing compliance for the majority of even the basic structures is imperfect, then how can building departments be expected to review the more complex structures? This paper will probably not stop the continued advancement of the complexity of the code, but it does suggest that the solution to the more complex structural work is independent peer review. Such review would open the door for the knowledgeable designer to use proven state-of-the-art knowledge in the analysis of complex buildings.

The lesson learned is that the more complex structures that do not fit into the category of simple regular buildings will never completely fit into code provisions, and at worst the code will be terribly messed up if the basics are obscured while attempts are made to cover all the exceptions. A better solution is to require independent peer review for the exceptional buildings. There needs to be a balance between prescriptive codes and performance-code requirements.

Detailed Inspection

While attention is focused on complex designs and soil-structure interaction, another key link in the performance reliability is overlooked. This link is quality control and, in particular, inspection. There were numerous reports of damage directly attributable to structural elements being either improperly installed or missing altogether. The range of these defects extends from residential structures to multi-story facilities. One may question how the inspector could have missed that. Clearly it is time to do better. There needs to be a national mandate to get a qualified inspector on site, properly paid, and with enough time to confirm that all the seismic elements are properly installed.

That simple owner-paid investment seems warranted if the owner expects to be eligible for relief funds should the building ever be damaged.

Unreinforced Masonry

It is no surprise that URM structures performed poorly. It is, however, instructive to note that there were no deaths to occupants of unreinforced structures. All the deaths associated with URM poor performance occurred outside of the structure, either on the sidewalk, street, or in an adjacent building. Whether or not this is an anomaly or not is not the issue. A URM building is a danger to more than just the occupants, it is a danger to the adjacent activities.

The Seismic Safety Commission vigorously discussed the value of posting hazardous buildings and, in particular, URM structures. The findings published in a Report on Posting of a hazardous building might have some impact on the persons wishing to enter the building, but how would the hazard be reduced for those adjacent to the URM structure?

The lesson is simple and well known; unreinforced masonry buildings are potentially hazardous. The ICBO Code for Building Conservation proposes standards for evaluating such buildings, but clearly the intent of that ordinance is for life safety only.

Need for Interaction Between Research and Practice

Returning to the relationship between research and design, there seems to be a tendency to demand that researchers immediately translate research into new design rules or code provisions. On the other side, researchers tend to be somewhat narrowly focused on the issue in front of them with less involvement on the design meaning on a broad scope. It would be wise for neither to become impatient with the other. The cycle is clear—research advances knowledge and must be seasoned with the testing of the design community in order to draw the applicable lessons for use in practice. This test should also be used in reverse to assist the research community in looking at research that will support the needed advancements in practice. The Earthquake Engineering Research Institute, SEAC, and the Seismic Safety Commission all have research committees that are trying to develop a national research strategy to deliver the most bang for the buck in terms of usable product.

Direction and Funding of Research

Given the parameters of effective earthquake mitigation, it seems appropriate to focus research in areas where it will be most effective in improving seismic hazard mitigation. The inventory of damage to buildings should provide direction on prioritizing research. If it is true that the majority of the building stock is not high rise or complex structures, then this stock of buildings should be a large focus of the research effort. Equally clear then would be the suggestion that funding for this majority would be in the best interest of the funding agencies working to mitigate the hazard. For those complex and special structures making up a minority of the stock, it is time to consider a mechanism by which those specialty projects carry their own burden of research needs. The Japanese model of research funding coming through construction of buildings to create the demand for research should be considered.

Ground Motion Data

The focus on the instrumental data from the Loma Prieta earthquake is credible, but the research effort needs to be directed toward results that will benefit the knowledge of the designers of buildings throughout the United States. To this end, the prediction of the spectral characteristics of sites throughout the country would be a direct and needed goal. The understanding of the energy input and how it varies throughout the country is needed as the profession starts to look at energy concepts for seismic design. However, of immediate use would be the correlation of predicted building response to recorded building response. This correlation needs to address the traditional use of elastic models with a magic word, "damping," thrown in to resolve the differences between actual and theoretical performance.

Attention to the unusual effects of site soils in moderate earthquakes seems warranted and has broad implications for areas of lower seismicity.

CLOSURE

The Loma Prieta earthquake certainly should be viewed as a wake-up call mandating seismic-mitigation activity throughout the state of California. Unfortunately, action wanes as time separates the reality of the event. Outside of California, there may be less of a sense of urgency, less of a sense that there is a risk. Each community in the country should at least take the step to understand what their real risk is and deal with the judgments necessary to determine what level of risk is acceptable. If there are non-believers, then their attention should be directed to the 27,000 damaged residences resulting from the Loma Prieta event and the deaths resulting from URM falling onto adjacent sites. Implementation of a mitigation program responsive to the decision on the acceptable level of risk starts with the understanding of the simple fundamentals and not with the adoption of prescriptive codes.

CITED REFERENCES

ATC. In press. ATC-33. Preparation of Guidelines for Seismic Rehabilitation of Buildings, ATC, Redwood City, California.

California Division of Mines and Geology. 1990. CSMIP Strong-Motion Records from the Santa Cruz Mountains, California Earthquake of October 17, 1989. Report OSMS 89-06. As presented in the Proceedings from the SEAC 59th Annual Convention.

California SSC. 1992. The Homeowner's Guide to Earthquake Safety. Sacramento, California, SSC 92-02, October.

California SSC. 1991a. Loma Prieta's Call to Action. Report on the Loma Prieta Earthquake of 1989, Sacramento, California.

California SSC. 1991b. Policy on Acceptable Levels of Earthquake Risk in State Buildings. Sacramento, California, SSC 91-1, January.

California SSC. 1991c. Breaking the Pattern, A Research and Development Plan to Improve Seismic Retrofit Practices for Government Buildings. SSC 91-05.

California SSC. 1992. Phase II Research and Development Action Plan, Proposition 122, Sacramento, California, October 7.

CALTRANS. 1991. Proceedings of the First Annual Seismic Research Workshop. California Dept. of Transportation, Sacramento, California.

City of Oakland Memorandum. 1990. Earthquake Damaged Buildings—10% Criteria. March 13 (revised April 4).

Fratessa, P. F., and F. Turner. 1991. The New Policy on Acceptable Levels of Earthquake Risk in State Buildings. In Proceedings of the 60th Annual Convention, Structural Engineers Association of California, Sacramento, California.

Freeman, S., J. A. Mahaney, T. F. Paret, and B. E. Kehoe. 1993. The Capacity Spectrum Method for Evaluating Structural Response During the Loma Prieta Earthquake. Preprinted for the 1993 National Earthquake conference, Memphis, Tennessee, May 9.

Holmes, W. T. 1990. San Francisco Unreinforced Masonry Buildings Study. In Proceedings of the 59th Annual Convention, Structural Engineers Association of California, Sacramento California.

Kasai, K., and B. F. Maison. 1990. Structural Pounding Damage Due to Loma Prieta Earthquake. In Proceedings of the 59th Annual Convention, Structural Engineers Association of California, Sacramento, California.

Lomax, A., and B.A. Bolt. 1992. Broadband Waveform Modeling of Anomalous Strong Group Motion in the Loma Prieta Earthquake Using Three-Dimensional Geological Structures. Geophysical Research Letters. Vol. 19, pp. 1963-1966, October 2.

McClure, F.E. 1991. Analysis of a Two-Story Oakland Office Building during the Loma Prieta Earthquake. SMIP91: Seminar on Seismological and Engineering Implications of Recent Strong-Motion Data. California Division of Mines and Geology, 13-1-13-11.

Roberts, J.E. 1990a. Putting the Pieces Together: The Loma Prieta Earthquake One Year Later. In Proceedings, Bay Area Regional Earthquake Preparedness Project, October 15-18, San Francisco, California. Bridge seismic research sponsored by CALTRANS.

Roberts, J.E. 1990b. Recent advances in seismic design and retrofit of bridges. In Fourth U.S. National Conference on Earthquake Engineering Proceedings. EERI, Palm Springs, California.

SEAC. 1991. Reflections On the Loma Prieta Earthquake October 17, 1989. Ad Hoc Earthquake Reconnaissance Committee, Chapter 9, Response of the Disaster Emergency Services Committees.

Webster's Intermediate Dictionary. 1977. Merriam-Webster Inc. Publishers, Springfield, Massachusetts.

Werner, S.D., A. Nisar, and J.L. Beck. 1992. Assessment of UBC Design Provisions Using Recorded Building Motions from Morgan Hill, Mount Lewis, and Loma Prieta Earthquakes. Dames and Moore, Oakland, California, April.

GENERAL REFERENCES

The following individuals and written material have been used in formulating this paper. The conclusions drawn from these references do not necessarily reflect the views of the individuals or authors listed.

Responding Professionals: Robert Burkett, Pat Campbell, Lloyd Cluff, Edward Diekmann, Neville Donovan, Robert Dyson, Eric Elsesser, Robert Englekirk, Sig Freeman, S.K. Ghosh, Mel Green, James Hill, Ronald Hamburger, William Holmes, George Housner, Wilfred Iwan, John Kariotis, James Libby, Charles Lindberg, Jack Meehan, William Menta, Jack Moehle, Rawn Nelson, Joe Nicoletti, Chris Rojahn, Mete Sozen, Don Strand, Tom Tobin, Ajit Virdee, Ted Zsutty.

DOCUMENTS

Ad Hoc Earthquake Reconnaissance Committee. 1991. Reflections On the Loma Prieta Earthquake October 17, 1989. Structural Engineers Association of California, Fair Oaks, California.

California SSC. 1991. Loma Prieta's Call to Action, Report on the Loma Prieta Earthquake of 1989. Sacramento, California.

California SSC. 1990. Report to Governor George Deukmejian in response to Executive Order D-86-90, November 30, 1990. Sacramento, California. Report SSC 90-06.

National Clearinghouse for Loma Prieta Earthquake Information:
Catalog April 1991
Catalog November 1991
Catalog April 1992
Catalog November 1992

National Information Service for Earthquake Engineering/Loma Prieta Project. Earthquake Engineering Research Center, 1302 South 46th Street, Richmond, California 94804-4698.

Proceedings, 59th Annual Convention Structural Engineers Association of California. 1990. Structural Engineers Association of California, P.O. Box 399, Fair Oaks, California 95628.

Proceedings, 60th Annual Convention Structural Engineers Association of California. 1991. Structural Engineers Association of California, P.O. Box 399, Fair Oaks, California 95628.

DISCUSSANTS' COMMENTS: BUILDINGS SESSION

James Beavers, Martin Marietta Energy Systems

The real lesson I've learned from the Loma Prieta earthquake, as an observer primarily, is that we need to know more about the performance of buildings during earthquakes, especially in moderate seismic zones. We've been focused on retrofit policy issues—how to understand performance, how to evaluate these facilities. As Paul Fratessa mentioned, only 2 percent of Loma Prieta research dollars was for unreinforced masonry. Yet workshops I have attended have cited the critical need to understand better the performance of unreinforced masonry, as well as non-ductile concrete.

We have been doing a major program on understanding the performance of hollow clay tile, a common wall type in Tennessee. Some have said hollow clay tile walls don't have ductility, but tests showed differently—significant capacity out of plane. One of the key lessons we've learned through our tests is to understand completely the performance from a structural engineering standpoint. How are these components going to perform during an earthquake? These are critical questions that must be addressed in areas such as east Tennessee.

We all know that equipment in buildings when properly restrained or anchored has exhibited satisfactory earthquake performance. Where failures do occur, they can be attributed to inadequate or missing anchorage. Good performance of equipment subjected to past earthquakes is not surprising considering that equipment failures observed during shake table testing have been rare. We believe that it seems quite apparent that rather than qualification testing of equipment to prove that it can withstand earthquakes—emphasis should really be placed on good fabrication, good assemblage practice, quality assurance, and inspection. Thank you.

Stephen Mahin, University of California, Berkeley

The Loma Prieta earthquake provided the first major opportunity in two decades for the earthquake engineering profession in the United States to assess its design procedures. While the region of intense ground motion was relatively remote, significant motions did develop in the vicinity of many major engineered structures. Especially important was the presence of numerous strong-motion instruments installed in buildings and at free-field sites.

The earthquake reemphasized many lessons from past earthquakes. These lessons include:

• the importance of soil conditions to the intensity of motion at a site and to structural damage;
• the significant damage to older (and some newer) residential construction, and the personal and social consequences of this damage;

- the need to retrofit structures, such as unreinforced masonry buildings, non-ductile reinforced concrete, and so on (many retrofit structures suffered significant structural damage under moderate shaking; studies to determine appropriate retrofit procedures are needed);
- the need to develop effective methods to assess the seismic resistance and structural integrity of damaged structures and to perform emergency as well as long-term repairs;
- life safety includes consideration of falling hazards as well as structural collapse;
- the need to develop methods for characterizing the importance of pounding of adjacent buildings and to mitigate the consequences of this pounding; and
- the increasing importance of functionality and repairability following a major event.

All of these items require additional research. In addition, the earthquake did not provide compelling evidence regarding the behavior of structures on soils that liquefy; the performance of buildings, nonstructural components, and contents when subjected to design-basis shaking; the performance of new construction types; and the efficacy of current retrofit and repair techniques.

More detailed quantitative investigations of the Loma Prieta earthquake have been aided by the numerous records obtained during the event. However, these records remain under utilized due to limitations of funds to investigate their structural implications. While detailed engineering studies have been carried out on most damaged structures, much of this data has not entered the public domain due to legal considerations and the unavailability of a suitable forum. Similarly, public access to structures and information regarding structures, particularly modern engineered structures, has been poor. Similarly, statistical summaries of this information in terms of failure/damage rates for different types of structures, down time, repair and disruption costs, and so on have not been systematically gathered. This information would be invaluable in assessing the performance of structures in other earthquakes in the Bay Area and elsewhere. While short-term reconnaissance style studies are carried out, few financial and personnel resources are available for sustained and directed engineering studies of these types.

The Loma Prieta earthquake also raises many questions. For example, what structural attributes allow some older buildings to remain undamaged, while nearby, newer structures are damaged; why nearby buildings of similar construction have dissimilar performance; how to design structures in a region where the intensity of motion may only be locally severe (microzonation for near field, soil amplification and ground hazard effects); and how to design structures to remain functional or reparable following major earthquakes. These and other questions call for concerted, focused efforts by practicing engineers, building-code officials, owners, policy makers, and researchers.

Gerald Jones, Kansas City Building Department

Perhaps the most difficult concept to explain to the average citizen is that of acceptable risk. Society readily accepts the deaths of 300 to 500 people each weekend on our highways in vehicle accidents, but has little tolerance for *any* loss of life in building-related incidents. The same citizens are the first to complain when new code requirements raise the cost of their proposed new buildings or the cost of maintaining an effective building department. One of the greatest challenges is defining "how safe is safe." When society reaches agreement on "how safe we should be," it is necessary to carefully describe the expectation level the code and regulations can deliver. Minimum legal levels must be differentiated from desired levels.

A building code is based on three primary issues—health, safety, and general welfare. Few people would disagree with the desire to cast codes in more performance-oriented formats. The issue becomes one of measurability—how does one know when the desired level of performance has been achieved without certain prescriptive yardsticks? This issue also raises the questions of competency for the designer, the builder, and the building official and of how the owner (the ultimate source of funding) can make the economics pay for the increased costs. The role of the building department is to act as the conscience of the community in the building process. I cannot agree more with Tom Tobin that there is a major difference between having a code and enforcing a code.

All in all, this symposium is an excellent opportunity to learn from each other. Thank you.

William Holmes, Rutherford & Chekene

We have heard many comments how, since the Loma Prieta earthquake, engineers have had to deal with policy makers or make policy themselves, much more so than ever before—such as how to define performance objectives and acceptable risk.

We need to settle some of these policy issues on where to put our mitigation resources. First, we need to look more closely at the assumed (non-engineering) deficiencies for residencies, and ask what the motive is for fixing these—life safety or economy. Water heaters should differentiate between gas and electric for life safety; the benefit in cost ratio is high. Chimneys are not life-threatening, and it costs as much to remove them as to repair the damage. The fixing of cripple studs and foundation bolt problems has a very high benefit in cost ratio. The more serious structural problems, such as side-hill sites and soft stories are life-safety hazard, but it is necessary to establish a standard to estimate a cost in benefit ratio and to mandate repair.

Second, we must look at how to decide when a damaged building can be reoccupied. Engineers have to identify the motive for repairing deficiencies in damaged buildings. The policy choices are (1) to repair buildings to a level where the re-occupancy risk is no greater than it was before the earthquake; (2) if the earthquake revealed an obvious hazard, to mitigate it before occupancy was allowed (repair building elements); (3) to require retrofitting for buildings shown to be vulnerable by the earthquake above some trigger level; and (4) to look at damage patterns, deciding which class of buildings are hazardous and passing ordinances requiring all buildings in that class to be fixed, even if not damaged.

Problems for engineers include the methods and effectiveness of repairing elements to their original conditions. What can be done? How can the trigger be set for repair versus retrofit requirements? It is difficult to identify the extent to which the building has really been damaged.

Lastly, the Loma Prieta earthquake brought to light a whole set of *adjacency hazards*. These include pounding where two buildings collide, vertical dynamic irregularities when a high building is adjacent to short one, local instabilities caused by unaligned floors of adjacent buildings, hazards caused by a shared wall, and components of one building falling on the other. As we have no control over adjacent properties, the best we can do is notify the owner that there is a potential problem and perhaps design the strengthening of the building with the adjacent hazard in mind.

To solve these problems, engineers must get the policy makers and lawyers involved. Thank you.

4

Emergency Preparedness and Response

Kathleen J. Tierney

INTRODUCTION

One of the most costly and damaging disasters in U.S. history, the 1989 Loma Prieta earthquake was the largest earthquake to strike California since 1952 and the most devastating to hit the San Francisco Bay area since 1906. From the earliest hours following impact, as initial reconnaissance efforts got under way, it was evident that the Loma Prieta earthquake would become an important case study for the various disciplines concerned with earthquake hazard reduction. Because the earthquake presented such an obvious opportunity to learn more about the earthquake hazard, an unprecedented number of studies were undertaken in the earth sciences, engineering, and the social sciences. By 1993, more than three years after the event, an enormous amount of data have been collected and much has been learned on a wide range of topics. This paper focuses on the lessons for emergency preparedness and response that have resulted from that research. After presenting a brief overview of research on the Loma Prieta earthquake, the paper considers research findings and practical implications related to the actions of individuals, households, and the public at large; groups and organizations; and government agencies and the intergovernmental system.

OVERVIEW OF PREPAREDNESS AND RESPONSE STUDIES

The Loma Prieta earthquake was the most damaging earthquake to strike a major metropolitan area in the United States since the passage of the National

Earthquake Hazard Reduction Act in 1977. One of the original objectives of this act was to foster needed research on earthquakes, and since the program was established the size of the research community in the earthquake field had grown considerably. Thus, when the Loma Prieta earthquake struck, a large number of investigators were able to go into the field almost immediately to begin collecting data, and ultimately dozens of studies were undertaken on a wide range of topics.

The Natural Hazards Research and Applications Information Center at the University of Colorado, which provides "quick response" grants mainly to social science investigators, was an important source of funding for initial reconnaissance studies on emergency preparedness and response following the earthquake. The center's monograph, *The Loma Prieta Earthquake: Studies of Short-Term Impacts* (Bolin, 1990), is a compilation of reports from nineteen investigators who conducted studies on the emergency response and initial impacts of the earthquake.

As it does for all significant earthquakes (but this time on a very large scale), the Earthquake Engineering Research Institute (EERI) undertook a major reconnaissance effort following the Loma Prieta earthquake, which was coordinated in the social sciences and emergency-response areas by Robert Olson and Charles Scawthorn. Reports of the various EERI reconnaissance teams are summarized in a special issue of *Earthquake Spectra* (EERI, 1990).

Following the earthquake, both the National Science Foundation and the U.S. Geological Survey augmented their existing grant programs to sponsor additional studies, some of which focused on emergency preparedness and response. Findings from much of that research are still in the process of being released as part of the U.S. Geological Survey Professional Paper Series. One collection of social scientific reports in this series, edited by Patricia Bolton, was recently published (Bolton, 1993). The National Center for Earthquake Engineering Research also provided support for reconnaissance studies, longer-term research, and other activities, including partial funding for a conference on post-earthquake housing issues that focused on the Loma Prieta earthquake as well as other recent U.S. earthquakes (National Center for Earthquake Engineering Research, 1992).

Because of the large number of researchers involved, the comparatively large amount of funding that was provided, and the size and sophistication of many of the studies that were undertaken, there are probably more data available on the Loma Prieta earthquake than on any other disaster. Efforts such as the National Clearinghouse for Loma Prieta Earthquake Information, organized by the National Information Service for Earthquake Engineering, help ensure that these data are preserved and used.

It would not be possible for this paper or any other short paper to discuss in detail all the reports on emergency preparedness and response that resulted from the Loma Prieta earthquake. Rather, the paper will highlight some of the more important findings and lessons learned, referring readers who desire more detail to the longer reports.

HOW THE PUBLIC RESPONDED IN LOMA PRIETA: INDIVIDUAL AND HOUSEHOLD RESPONSES

Several studies on the Loma Prieta earthquake provide useful data on how the public responded when the earthquake struck and during the post-impact emergency period. Among the most important of these are studies on the initial post-impact actions of community residents, earthquake-related injuries, emergency sheltering behavior, and the public response to aftershock warnings.

Actions During the Shaking Period

Using a survey approach like the one they employed after the 1987 Whittier Narrows earthquake, Linda Bourque, James Goltz, and their associates conducted a telephone survey with a random sample of 656 respondents in the five counties most seriously affected by the Loma Prieta earthquake. The survey focused on a number of topics, including what people did during and immediately after earthquake impact, property damage and injuries, decision making with respect to evacuation, psychological distress at the time of the interview, earthquake preparedness actions taken before and after the earthquake, reliance on the mass media after impact, exposure to aftershock warnings, and contacts with public agencies following the earthquake (Bourque et al., 1993a, b).

With respect to immediate actions upon impact, the survey found that the most common responses during the shaking, carried out by nearly three-fourths of respondents, were to freeze in place, to seek protection, or to freeze and then seek protection. The researchers observed that workplace disaster-preparedness programs must be having an effect because a significant proportion of those who were at work or in schools at the time of the earthquake reported taking self-protective actions during the shaking.

Research conducted by Rahimi and Azevedo (1993) of a sample of disabled persons following the earthquake suggests that, like the individuals in the Bourque/Goltz sample, people with disabilities were able to initiate appropriate actions to protect themselves during earthquake shaking. They may, however, be less able than the non-disabled to gain access to personal items and emergency supplies after earthquake impact.

Running during earthquake shaking is not considered appropriate, because it may result in injury. A small number of respondents in the Bourque/Goltz survey indicated they ran at the time of impact; running was more likely to be reported nearer the earthquake epicenter and more likely to be reported by young males, indicating that this group may be less aware that such behavior is dangerous. About 40 percent of the people who were at home when the earthquake struck reported going to help a child during the shaking period—an action that may have increased personal risk.

This study on individual post-impact responses revealed other interesting patterns. For example, persons who were not at home, work, or in some other familiar setting but who instead found themselves in public places when the earthquake struck seemed less able than other respondents to take decisive self-protective action. Because public places are less familiar than the home or workplace, people may be confused about what to do to protect themselves in those settings. On the basis of this finding, the researchers suggested that:

> Procedures need to be developed that allow people to identify and act on "generic" information about locations which can then be generalized from one location to another. In particular, people need to know how to scan and quickly assess a location for safety, and how to behave in the presence of large numbers of other people so that they do not endanger themselves and others.
>
>   Bourque et al., 1993b

For more detailed discussions of these survey findings, see Bourque et al. (1991, 1993a, b).

John Archea (1990) interviewed 41 Santa Cruz residents who were in their homes at the time of the earthquake, obtaining detailed reconstructions of what people did during the shaking period. His preliminary data indicate that during the 10- to 12-second period of strong shaking, building occupants engaged in a variety of actions, including seeking refuge from moving and falling objects, simply staying as still as possible and "riding out" the shaking (likely equivalent to the "freezing in place" described in the Bourque/Goltz survey), attempting to protect property by bracing or propping it up, and trying to go outside. A few respondents were unable to initiate any action at all during the shaking. Archea observes that while the majority of those interviewed actively tried to protect themselves, "they also unwittingly took some risks to do so, and many were still at risk 10 or 12 seconds later when the shaking stopped."

Evidence from the Loma Prieta earthquake suggests that Bay Area residents were aware of what to do when an earthquake strikes and that even during the strong shaking period they were capable of making choices and taking actions to decrease their vulnerability. Public education programs are paying off. At the same time, the programs aren't reaching everyone who needs them, and some people continue to take unnecessary risks.

Injuries

Reports on the distribution of various types of injuries differ somewhat due to differences in the data sources and classification systems used. Dr. Tierney (1991) reports that approximately 1,100 persons were seen in hospitals in the six-county area of impact for earthquake-related injuries and medical complaints on the night of the earthquake; of this number, 73 percent were treated and released. Most of the injuries were not severe. Wounds, abrasions and contusions, fractures, and sprains and strains were the most commonly treated injuries, accounting for more than half of the total. Durkin et al. (1991), using data from the California Department of Industrial Relations and the Office of Emergency Services, found a similar pattern: strains, sprains, and contusions constituted 60 percent of the shaking-related injuries and 70 percent of the post-shaking injuries.

The survey conducted by Bourque, Goltz, and their colleagues after the earthquake sought data on the entire population in the affected area, rather than only persons who visited hospital emergency departments or made injury claims. In that study, a very small number of people in the randomly selected sample— 7 out of the 656 persons surveyed—reported having been injured in the earthquake. The likelihood of being injured was highest for residents of the hardest-hit areas of Santa Cruz County; about 3.3 percent of those surveyed in those areas reported having been injured (Bourque et al., 1993b). These rates of reported injury differ from those found by some investigators, as is discussed in more detail below. However, they are consistent with injury rates these same investigators found following the Whittier Narrows earthquake.[1] In the Whittier Narrows event, the injury rate was 26 per 1,000 residents in the area of severest earthquake shaking; in the Loma Prieta earthquake, the rate was 32.8 per 1,000 in the hardest-hit area of Santa Cruz County (Bourque et al., 1993b).

After the earthquake, a multidisciplinary team of researchers began collecting data on Santa Cruz County residents who died and who were treated at hospitals for earthquake-related injuries, as well as on a randomly selected control group matched by area of residence. The objective of the study was to identify risk factors for death and injury, as well as for injury severity. Based on earlier studies, the researchers hypothesized that such risk factors would include various attributes of the physical environment in which the individual was located at the time of the earthquake, such as building type; the behaviors undertaken by the individual, such as self-protective actions; whether the individual was able to move at the time of impact; sociodemographic characteristics of the individual, such as age; aspects of the setting at the time of impact, such as whether the individual was alone or with others; and other risk factors, such as whether the individual had pre-existing medical problems or disabilities. Data collection on this study is complete; data were obtained on 483 persons in the injured/killed sample and 701 persons in the control group. (For detailed discussions of analyses that have been conducted to date, see Jones et al., 1992, 1993 and Wagner et al., 1993).

---

[1] That is, the injury rates differ moderately, but in the expected direction. Loma Prieta was a larger earthquake, covering more area, with a longer period of strong shaking. Soil conditions amplified the ground shaking in some areas, which doubtless contributed to the incidence of injuries during the shaking period.

Detailed findings are not yet available on specific risk factors, but this study has already provided some very useful information. For example, the data on when deaths and injuries occurred indicate that while the majority happened at the time of earthquake shaking, a rather high proportion—about 40 percent—occurred within the 72-hour time period after the earthquake, suggesting it may have been possible to prevent some of these later injuries, for example by issuing advisory warnings to the public. In conducting the population survey to obtain the "non-injured" control group, the investigators also found that 15 percent of those contacted reported actually having been injured in the earthquake, even though they did not seek hospital treatment for their injuries. A similar pattern is discussed in reports by Durkin et al. (1991) and Thiel et al. (1992), which indicate that as many as 60 percent of those with earthquake-related injuries either treated themselves or received treatment in non-hospital settings. These findings suggest that a substantial number of earthquake-related complaints were dealt with outside the formal health care system. They also raise the question why some injured persons elected to seek hospital treatment while others did not.

However, as noted earlier, the findings from studies on overall numbers and rates of injuries resulting from the Loma Prieta earthquake are not consistent. It is unclear why one survey (Bourque et al., 1993b) found that roughly 3 percent of respondents in the most severely affected communities in Santa Cruz County were injured, while another (Jones et al., 1992) suggested that the county-wide percentage may have exceeded 15 percent. Such discrepancies could be due to a number of factors: the wording of questions about injuries, how soon after the earthquake the surveys were conducted, the time period covered by the questions (e.g., during and immediately after the earthquake or after a longer period of time), the context within which the survey questions were asked, or the approaches used in establishing sampling frames and selecting samples.

With respect to the risk factors examined to date, data from the Santa Cruz injury study conducted by Jones and his collaborators indicate that being inside a building rather than outside when earthquake shaking began meant that a person had a 3.3 times greater risk of being injured. Studies using other data sets also reveal some intriguing patterns related to injury risk. Durkin et al. (1991) and Durkin and Thiel (1992), focusing on 18 earthquake fatalities and 325 injuries defined as "work-related," found that all but one of the fatalities were due to some form of structural failure.[2] Of the injuries occurring during earthquake shaking that were not related to structural collapse, 26 percent were attributable to falls (particularly stairway falls), and 21 percent occurred when people were thrown against objects. Falling and overturning objects accounted for another 28 percent of the injuries. These researchers also found that taking protective

---

[2] Forty-two of the 62 deaths attributed to the earthquake occurred in the collapse of a single structure—the Cypress freeway. The other major cause of death was the total or partial collapse of unreinforced masonry buildings.

action (getting under a desk, standing in a doorway) was sometimes associated with injury, but those injuries tended to be minor. They suggest that while the recommended self-protective actions may enhance life safety in collapse-hazard situations, people who rush to protect themselves in other less hazardous settings may be increasing their risk of minor injury.[3]

Evacuation and Use of Emergency Shelter

The earthquake caused many residents of the Bay Area to vacate their homes, either temporarily or permanently, and to seek various forms of emergency shelter. Bourque et al. (1993a, b) found that overall 22 percent of the respondents in their survey reported having evacuated for at least some period of time. Propensity to evacuate varied according to the severity of earthquake effects, with the highest proportion of residents (about 43 percent) evacuating in the heavily damaged areas of Santa Cruz County. Most of the respondents in this sample returned to their homes within 24 hours.

The study found that the tendency to evacuate was higher nearer the earthquake's epicenter and higher for respondents whose homes suffered damage. However, an important lesson this earthquake brought out is that physical earthquake effects such as damage and loss of utilities were by no means the sole factor explaining evacuation. Many people with such damage did not evacuate. A substantial proportion of those that did evacuate (up to half of the respondents in the five-county survey) reported doing so for reasons that were unrelated to physical damage levels, such as emotional upset, fear of aftershocks and further damage, and concern about the safety of their children.

As is the case in most disasters, the majority of evacuees made their own sheltering arrangements after the earthquake, mainly staying with friends and relatives or camping outside near their homes. Regarding "official" shelter use, Bolin and Stanford (1990) report that at the peak period 2,500 displaced persons were being provided with shelter nightly, about 20 percent of the estimated 12,000 to 13,000 left homeless. By the end of the third week after the earthquake, all but about 500 of those displaced were either relocated into temporary housing or were back in their homes.

Studies of the provision of temporary shelter after the Loma Prieta earthquake have revealed some interesting patterns with important policy implications. First, the research clearly shows that post-earthquake temporary shelter needs are closely related to a community's pre-earthquake housing problems.

---

[3] Findings like these should be considered preliminary and suggestive, however. Research on how people get killed and injured in earthquakes is still in a very rudimentary stage, and much of the work done to date has been methodologically flawed. When the data from the Loma Prieta earthquake are analyzed fully, more information will be available on who got injured and how than exist for any other earthquake. Nevertheless, many more studies will be needed before it will be possible for researchers to develop adequate explanations for earthquake-related mortality and morbidity.

For example, in the Loma Prieta earthquake, those most likely to be displaced were low-income persons, usually tenants, who had inhabited older, low-rent properties before the earthquake—properties that were already in poor condition and very short supply. Those people also needed to remain in shelters longer than is usually the case in U.S. disasters, because with so many units destroyed and uninhabitable, it was even more difficult after the earthquake for them to find suitable housing.[4] Homelessness was already a problem in the Bay Area before the earthquake. The Loma Prieta earthquake damaged homeless shelters and a number of the single-room-occupancy hotels that are an important source of housing for the very poor, and as a result homeless people and those at risk of becoming homeless were even worse off after the earthquake than before.

Second, although disaster assistance agencies may intend shelters to be used by disaster victims only, definitions of who constitutes a "victim" may differ. For example, in some communities affected by the earthquake, because there was already a significant homeless population and a shortage of housing, "pre-disaster homeless" persons attempted to utilize shelter facilities and other services. Some of these individuals did not meet the official eligibility requirements for disaster assistance. Although it was the position of the agencies that disaster aid was not meant to address what they considered pre-existing community problems, community groups argued that programs should try to meet the needs of everyone affected by a disaster, rather than defining eligibility in strict bureaucratic terms.[5]

Third, the Loma Prieta earthquake showed clearly that as the major metropolitan areas of the United States become increasingly ethnically and racially diverse as a consequence of immigration and other population trends, the population requiring post-disaster sheltering and other services will reflect that diversity. It will be necessary to tailor assistance programs to the needs of program users rather than delivering "generic" services in a standardized manner. (For more detailed discussions of housing issues, see Bolin and Stanford, 1990; Phillips, 1991; Phillips and Hutchins, 1991; and Bolin, 1993.)

Response to Aftershock Warnings

Following the M7.1 mainshock, numerous aftershocks occurred, and aftershock warnings continued to be issued to the public over a two-month period.

---

[4] One of the Santa Cruz County shelters remained open for sixty-six days after the earthquake—an extremely long period for a "temporary" shelter (Phillips, 1991).

[5] Following the Loma Prieta earthquake, the Federal Emergency Management Agency (FEMA) was severely criticized for disaster assistance policies that were believed to discriminate against low-income households, members of the homeless population, and people in transient living situations (GAO, 1991). A class action lawsuit filed against FEMA in 1989 resulted in an out-of-court settlement earmarking FEMA funds specifically for the reconstruction of housing for low-income Bay Area tenants.

Dennis Mileti and Paul O'Brien studied how residents of San Francisco and Santa Cruz counties responded to these aftershock warnings. They found that most people were aware of the aftershock warnings, particularly in Santa Cruz County, and many respondents (66 percent in San Francisco County and 75 percent in Santa Cruz County) believed damaging aftershocks would occur. By two months after the earthquake, substantial numbers of people had taken one or more additional preparedness measures, such as protecting household items from damage, and again this tendency was more pronounced in Santa Cruz County. However, the people most likely to pay attention to and act on aftershock warnings were those who had experienced damage in the mainshock and who subsequently got involved in the community emergency response. People who weren't affected by the mainshock tended to do less in response to aftershock warnings, leading the researchers to conclude that:

> Those who experience little or no loss in the impact of a disaster may be prone to a "normalization bias" when interpreting post-impact warnings for subsequent risk: "the first impact did not effect me negatively, therefore, subsequent impacts will also avoid me."
>
> Mileti and O'Brien, 1992

Such a conclusion would, of course, be unwarranted, and future aftershock warning efforts should emphasize that point. (For further discussions, see O'Brien and Mileti, 1992; Mileti and O'Brien, 1992, 1993.)

RESPONSE OF GROUPS, ORGANIZATIONS, AND INTERORGANIZATIONAL NETWORKS

A disaster of the magnitude of the Loma Prieta earthquake causes the mobilization of a vast range of organizational and community resources; it is probably fair to say that at the peak of the emergency thousands of different organizations and tens of thousands of people were involved in the response. Obviously, it was not possible for researchers to collect data systematically on all the individuals and organizations that performed disaster-related tasks. However, certain aspects of the response were well-studied, and a considerable amount was learned about how some types of organizations handled disaster-generated demands and coordinated their activities. This section focuses on some of the more important groups, organizations, and networks that were involved in the emergency response, including emergent groups and volunteers, local first-response agencies, hospitals and emergency health-care providers, transportation networks, utility lifeline organizations, and governmental emergency-response agencies.

Emergent Groups and Volunteers

Researchers have long been aware that disasters are invariably accompanied by an increase in altruistic or pro-social behavior on the part of the public (Barton, 1970; Dynes, 1970; Drabek, 1986). The Loma Prieta earthquake was certainly no exception to this pattern. A survey conducted by O'Brien and Mileti (1993) with a representative sample of residents in Santa Cruz and San Francisco counties found that a large majority of residents in both counties—70 percent in Santa Cruz and 60 percent in San Francisco County—participated in some type of emergency response activity following the earthquake. Among the most widely reported activities were providing food and water for others (35 percent in Santa Cruz, 14 percent in San Francisco), helping with cleanup and debris removal (44 percent in Santa Cruz, 11 percent in San Francisco), providing shelter to others (18 percent in Santa Cruz, 12 percent in San Francisco); and providing counseling to victims (17 percent in Santa Cruz, 8 percent in San Francisco). Three percent of San Francisco respondents and about 5 percent of Santa Cruz respondents reported engaging in emergency search-and-rescue activities. Although these percentages seem small, when extrapolated to the entire population of those counties, they add up to more than 31,000 search and rescue volunteers. Clearly, the response by the public was massive following the Loma Prieta earthquake, and a large share of the assistance that was provided to victims was given through informal channels.

Groups composed wholly or partly of volunteers were a critical element in the emergency-response effort. Such groups ranged from organizations with formal, long-standing disaster responsibilities like the Red Cross to newly formed building-damage inspection crews and emergent search-and-rescue groups. Focusing specifically on the initial response to the Cypress structure collapse, Garcia et al. (1993) note that community residents were massively involved from the earliest moments after impact, rescuing trapped motorists, helping victims to safety, and giving first aid. These volunteers provided assistance that was desperately needed, often putting themselves at risk to do so. However, because the volunteer response was so large, coordination of volunteers became a major challenge.

No comprehensive research has been conducted on the formation, activities, and effectiveness of volunteer groups in the emergency-response period. However, Neal (1990a, b) was able to examine the activities of nine volunteer organizations in one community following the Loma Prieta earthquake. These organizations included the Red Cross, the Salvation Army, an amateur radio group, and a local volunteer coordinating council. Among Neal's conclusions were that volunteer organizations were relatively effective even though their activities were often not well-coordinated with those of local government. Effective task performance for this group of volunteer organizations was related to (1) the degree of prior disaster planning, (2) the degree of prior disaster experience, and (3) the degree to which the organization had established ties with other important community organizations before the earthquake.

Local Emergency-Response Activities

The mobilization of emergency resources following the Loma Prieta earthquake was massive, and, in most cases, the resources available far exceeded the actual demand. As the discussions below will indicate in more detail, except in those communities near the epicenter, such as Santa Cruz and Watsonville, the emergency-response system was not strained during the event, even during the period of peak demand on the night of the earthquake. However, the disruption and demands produced by the earthquake were sufficient to suggest where problems might develop in future earthquakes. Key emergency-response tasks are discussed briefly below.

Search and Rescue

No major analytic reports on search-and-rescue activities following the Loma Prieta earthquake have been released; the accounts published to date are mainly descriptive. However, based on existing reports, it appears that except for the large-scale organized effort that developed at the Cypress structure and smaller formal rescue actions undertaken in the Marina District in San Francisco and the Pacific Garden Mall in Santa Cruz, the majority of the search-and-rescue that took place following the earthquake was conducted informally by community residents. And even in the more formally organized efforts, residents of the damaged areas and other volunteers played a major role.

Descriptions of the extensive search-and-rescue operation that was undertaken at the Cypress structure can be found in the city of Oakland's *Loma Prieta Earthquake After Action Report* (1990), EERI's Reconnaissance Report (1990), and a report on the activities of the Oakland Fire Department by Garcia et al. (1993). Those documents make the following observations about that search-and-rescue effort:

• There were major difficulties with interorganizational communication, because so many different responding agencies with different radio frequencies were involved.

• Convergence of personnel, vehicles, and equipment made management and coordination of search-and-rescue difficult at times.

• Convergence by the mass media also created problems.

• An overall Incident Command System was difficult to institute, because so many different agencies responded, and many responders were either unfamiliar with the system or used different versions.

• Rescue operations were hindered initially because of the lack of heavy rescue equipment and portable lighting.

• Community residents, ranging from persons living in the immediate vicinity to contractors and other individuals with specialized equipment, volunteered in the search-and-rescue effort in large numbers. Their contributions were extremely valuable, but because of the sheer number of people wanting to provide assistance, coordination problems did occur.

Emergency Medical Services (EMS)

Hospitals, ambulance companies, EMS agencies, and other emergency medical care providers were the subject of considerable study following the Loma Prieta earthquake. Despite the fact that several components of the EMS system were damaged and disrupted by the earthquake, overall response capacity was not compromised. Major damage was confined to a small number of hospital facilities near the epicenter; Watsonville Community Hospital was particularly hard hit but remained functional (for detailed information on damage, see EERI, 1990). Nonstructural damage to hospitals was widespread in the area of impact, but the damage by and large was not severe enough to interfere with patient care. Of the various components of the emergency medical care system, communications facilities such as the "911" dispatching centers were perhaps the most seriously affected by the earthquake. Difficulties with EMS communications and dispatching stemmed from a variety of sources: earthquake-induced power failures; damage to the buildings in which the facilities were housed; damage to critical equipment, such as computers; loss of computer-aided dispatching capability; disruptions in communications (particularly phone communications) between the centers and the outside community; damage to message-transmission facilities; and excess radio traffic (EERI, 1990; Tierney, 1991).

In their study on the operations of the "911" communications center in Santa Cruz County, Durkin et al. (1991) found that while the volume of calls was much higher than normal on the night of the earthquake and record-keeping suffered as a result, EMS personnel were able to respond to all requests for assistance. When dispatched into the field that night, emergency workers adjusted to the increased demand in several ways:

- several patients were treated and released by the EMS crews;
- multiple individuals were transported in one run to a hospital;
- patients were directed to other means of transport if available; and
- no action was taken when the crew determined that their services were not critical.

There was an unusually high number of cases where the ambulance arrived and found that other medical resources were already on the scene (Durkin et al., 1991).

The victims transported by EMS personnel during the emergency period were less seriously injured than those normally seen, requests for emergency assistance were extremely high only on the night of the earthquake, and the demand for EMS services dropped back to normal levels within about three days.

Ambulance personnel were available in very large numbers throughout the Bay Area on the night of the earthquake; in fact, there was an oversupply of emergency vehicles, paramedics, and emergency medical technicians in most of the damaged areas. Off-duty staff reported to area hospitals immediately following impact, and hospital personnel who were surveyed after the earthquake indicated that personnel and resources were more than adequate to deal with the medical problems that were seen (Tierney, 1991; Pointer et al., 1992).

The emergency medical care system functioned well in the earthquake, largely because (1) system capability was quite high in the area affected by the earthquake, as indicated by the quantity and quality of EMS resources; (2) essential health care resources survived the earthquake well, while flexible and redundant system components compensated for damaged and disrupted elements; and (3) the earthquake produced a comparatively small number of casualties relative to system capability, and most of the medical complaints that resulted were not severe (Tierney, 1991). However, the earthquake also showed there is a need to improve the ability of EMS systems to handle greatly increased numbers of calls, to determine rapidly which of those callers have the greatest need for specialized emergency services, and to allocate resources to the areas of greatest need (Thiel et al., 1992). Many of those who went to hospitals and requested emergency assistance from other EMS providers did not have injuries and medical problems severe enough to require those specialized resources. More attention needs to be paid to how large numbers of minor injuries will be handled in future earthquake events, particularly catastrophic earthquakes. Such efforts will obviously have to involve educating the public not to use emergency resources for non-emergencies.

Damage Assessment

Early Identification of Problem Areas. Early efforts to identify the areas that had been hardest hit in the Loma Prieta earthquake were complicated by the fact that communications were sporadic and the information available to response agencies was incomplete. Initial media reports greatly overestimated the number killed and injured, and by focusing on dramatic instances of damage the media (particularly television) gave the impression that the earthquake produced widespread destruction. The reports highlighted the damage in media-accessible areas such as the Marina District; underrepresented the losses in smaller, more remote areas such as Santa Cruz and Watsonville; and failed to put the amount of damage that had been done into perspective (Rogers et al., 1990).

Since media reporting is known to be selective, it is not advisable to rely on mass media reports of disaster impacts. However, in the absence of solid data, the initial impressions that were formed about the extent and location of the damage following the Loma Prieta earthquake—both by the general public and by emergency responders—were heavily influenced by media accounts. For example, local officials in Santa Cruz County, seeing the Marina District fires and the Bay Bridge damage on television, assumed the devastation was so widespread it was pointless to ask for resources from other counties (City of Watsonville, 1990).

Evaluation of Damaged Buildings. Systematic damage assessment to identify potentially unsafe structures began very soon after impact. Loma Prieta was the first earthquake in which the Applied Technology Council's *Procedure for Post-Earthquake Safety Evaluation of Buildings* (ATC-20) was used on a broad scale for this purpose.[6] Oaks (1990), focusing on how the evaluation process worked in San Francisco following the earthquake, makes several observations. First, aftershocks complicated the damage-assessment process, often making multiple inspections necessary; most red- and yellow-tagged buildings were reinspected an average of four times during the first week. Second, the damage caused by the earthquake also exposed asbestos in many of the buildings, which engendered controversy. Third, because the cost of evaluations must be borne by the property owner and decisions about what to do with buildings take time to make, the evaluation process for some damaged structures tended to get drawn out over time. Fourth, the damage-assessment process sparked landlord-tenant disputes, for example, when tenants were unwilling or unable to reoccupy buildings or when landlords used the earthquake as an occasion to evict tenants. Fifth, ATC-20 focuses on evaluating buildings but contains no directives for organizing and managing that evaluation effort—a monumental undertaking in a major earthquake.[7]

The damage assessment process and uncertainty about the safety of structures contributed to the ongoing need for shelter:

> Because of the ever-changing conditions, great resources were required in terms of time and personnel to carry out the reinspections and reassessments. The changing conditions also contributed to many social, economic, and legal problems that occurred as people were unable to continue to live in their homes or pursue their means of livelihood. For example, until buildings were inspected and considered safe for occupancy, it was uncertain if people could reinhabit certain structures. As a result, neighborhood and city resources faced demands for emergency sheltering.
>
> EERI, 1990

Oaks (1990) notes that despite these difficulties damage assessment activities in San Francisco went quite smoothly. This was due in large measure to the fact that there were so many trained, qualified persons ready to work as inspec-

---

[6] The evaluation guidelines had been published only a short time before the earthquake, and very few people had received training in their use.

[7] This comment was not meant to imply that ATC-20 should have done so—that was not the intention of the guidelines—but rather to suggest that organizing damage assessment is a significant task in and of itself, separate from the technical decision making that goes into the evaluation of buildings.

tors and because of the involvement of organizations like the Office of Emergency Services Volunteer Engineer Program and the California Association of Building Officials. Other evaluations of the ATC-20 process (e.g., California SSC, 1991) were also positive. The Loma Prieta earthquake clearly showed that the management of damage-assessment activities is a critical task in the emergency-response period. However, it also raises the question of whether other communities would be able to handle the task as well as those in the Bay Area.

Response of Public Transportation Networks

The collapse of the Cypress structure and the closing of the Bay Bridge and two other major San Francisco highways due to earthquake damage seriously disrupted transportation patterns in the Bay Area. San Francisco, Oakland, and the Bay Area in general faced the possibility of spiraling economic losses unless alternative modes of transportation could be developed to compensate for the loss of these key routes. Bay Area transportation agencies had not been involved in earthquake-preparedness planning to any great degree prior to the Loma Prieta earthquake. After the earthquake, these agencies became involved in intensive efforts to devise new transportation modes and routes that would bypass damaged connections in the system. Several hundred individuals and more than a dozen transportation agencies (including Bay Area Rapid Transit, Alameda/Contra Costa Transit, San Mateo County Transit, San Francisco Municipal Bus, the Golden Gate Bridge and Transit Services, and private ferry companies) participated in this effort. The system of transportation that was developed following the earthquake was especially critical during the first month after the earthquake, when the Bay Bridge remained closed. The existence of coordinating agencies like the Metropolitan Transportation Commission, the Transit Operators Coordinating Council, and the Regional Transit Association, as well as the fact that the various transportation agencies had a history of working together before the earthquake, helped the improvised system to get organized rapidly.

Lifeline Organizations

The Loma Prieta earthquake was in many respects a "lifeline disaster." Among the most dramatic examples of lifeline impacts were the collapse of the Cypress structure; the closures of the Bay Bridge, major highways in San Francisco, and Highway 17 due to damage; the loss of water for firefighting in the Marina District; and the damage to the Moss Landing electric-power substation. The major lifeline organizations in the Bay Area are highly prepared for disasters, particularly earthquakes. Many lifeline organizations had also been engaged in earthquake-hazard-mitigation activities prior to the earthquake that reduced damage and disruption. The high level of emergency preparedness was the principle reason why it was possible to restore lifeline services so rapidly after the earthquake. Large lifeline service providers like Pacific Gas and Electric documented their emergency-response activities extensively and produced after-action reports discussing lessons learned in the earthquake (see, for example, Phillips and Virostek, 1990). These reports are an important resource for both utilities and local governments seeking to improve their response capability.

Chapter 5 focuses on lifelines in depth, so they will not be discussed in detail here. One lesson that does warrant emphasis, however, is that many emergency-response activities depend upon lifeline systems in order to be effective, and lifeline damage can thus seriously impact community response capability. Isenberg (1992), for example, has documented the ways damage to lifelines affected emergency-response capacity in Watsonville after the Loma Prieta earthquake. Because of the loss of electrical power, for example, the city's emergency communications center could not function; it was difficult to pump gasoline from underground tanks; in order to get power, extensive use had to be made of emergency generators, which created additional problems; and hospital operations were adversely affected.

A related lesson from the Loma Prieta earthquake involves the extent to which the various lifeline services are interdependent. Electrical power is perhaps the most crucial service, because so many other lifelines need power in order to operate. Because lifeline services are so important to the overall community response, it is critical that linkages be maintained between lifeline organizations and local community officials for both pre-disaster preparedness and post-disaster response. Similarly, lifeline interdependence requires that the organizations providing different lifeline services engage in joint preparedness planning and coordinate their emergency activities.[8]

Local and State Emergency-Response Agencies

The earthquake did damage over an 8,000-square-kilometer area with a population of about 4 million people; six counties and dozens of local jurisdictions were affected. Except for large-scale multijurisdictional efforts like the Cypress structure search and the activation of some mutual aid agreements, jurisdictions generally handled their own emergency-response problems without much outside assistance. No systematic research was conducted on the effectiveness of local emergency-management systems following the Loma Prieta earthquake. However, many jurisdictions and organizations developed their own assessments of how well emergency tasks were performed and outlined the lessons they had learned (County of Santa Cruz, 1990; City of Watsonville, 1990; City of Oak-

---

[8] Such coordination was easier in the Bay Area following the Loma Prieta earthquake than it would be in many other communities, because Pacific Gas and Electric provides natural gas and power to the entire Northern California region. More commonly, the various lifeline services are provided by separate organizational entities, which hampers integration in planning and response activities (Tierney, 1992).

land, 1990). Hearings conducted by the California Seismic Safety Commission in affected jurisdictions also focused on emergency-response issues and problems (see California SSC, 1991).

The report of the State/Federal Hazard Mitigation Survey Team (1990) identified the following response-related needs that were highlighted by the Loma Prieta earthquake:

* formal procedures for the federal response to a major but not catastrophic earthquake;
* more-specific planning to assign responsibility for all Emergency Support Functions in the Federal Response Plan;
* policies and criteria to enable federal and state agencies to provide automatic assistance to local jurisdictions for time-critical response elements;
* a model emergency-management structure and procedures, common to local, state, and federal response agencies;
* enhanced communications systems at the federal, state, and local levels;
* a systematic approach to collecting data on damage;
* a model resource-tracking system for state and local jurisdictions;
* the identification of staging areas for various resources;
* the establishment of regional planning groups (e.g., in the Bay Area) to address response-related issues of regional concern;
* emergency medical service mutual aid agreements for the Office of Emergency Services regions, and a mutual aid plan for the provision of emergency fuel;
* efforts to address regional emergency transportation planning;
* increases in a broad range of emergency resources: generators, fuel supplies, search-and-rescue equipment, medical supplies;
* lists of federal, state, and local personnel who are capable of performing post-earthquake building inspections;
* increased capacity to provide short-term shelter to earthquake victims; and
* increased capacity to provide timely public information in earthquake situations.

The Seismic Safety Commission's report *Loma Prieta's Call to Action* (California SSC, 1991), which was developed with input from officials in several hard-hit communities, makes a number of observations and recommendations in the area of emergency response that warrant mention here:

* Local capabilities were sufficient to meet emergency-response needs, but this does not mean the Bay Area is ready for a larger earthquake.
* Earthquake-preparedness planning, disaster drills, and related activities helped local jurisdictions respond more effectively.
* The lack of accurate information about which areas were most severely

affected hampered the emergency response in the early hours and made some jurisdictions hesitant to request outside resources.

• A significant number of law enforcement, medical, and fire resources were provided through mutual aid agreements, and these arrangements generally worked well.

• The State Office of Emergency Services should be authorized to send resources to areas impacted by an earthquake automatically, regardless of whether local jurisdictions make specific requests.

• Guidance is needed on how to manage post-earthquake damage-assessment activities.

• California's emergency management system should be expanded and improved, and "a standardized Emergency Management System" should be developed for all governmental emergency organizations.

• Training of emergency managers and responders needs greater emphasis.

Although areas needing improvement are noted, the tone of the report is positive. It emphasizes the continuing need to overcome well-known barriers to better emergency management: budget shortages, the uneven quality of local emergency-management resources (e.g., emergency operations centers, communications equipment) and personnel, the use of inconsistent planning frameworks across local jurisdictions, limited state authority to mandate preparedness activities, and cumbersome rules about requesting and providing resources in disasters.

In light of what occurred in Florida following Hurricane Andrew, it is interesting to note that both the Seismic Safety Commission report and the Hazard Mitigation Team report stress the need for mechanisms to ensure the automatic provision of aid, bypassing the requirement that local jurisdictions (or states) must formally request resources from higher governmental levels in emergency situations.

In July of 1991, the Bay Area Regional Earthquake Preparedness Project convened a symposium to bring together officials from the cities of San Francisco, Oakland, and Los Angeles to discuss lessons the Bay Area cities learned from the Loma Prieta earthquake and to determine whether Los Angeles's planning assumptions needed to be modified on the basis of the Loma Prieta experience. The conference focused on five main areas of concern: managing the disaster response, issues related to public works, emergency shelter and housing, financial issues, and community and business preparedness. The report on the joint symposium (Governor's Office of Emergency Services, 1992) contains dozens of findings and specific recommendations on such topics as the management of human resources in the disaster-response period, coordination with the mass media, damage assessment and control of access to damaged sites, interagency coordination, debris removal, demolition, code changes, emergency shelter and housing recovery policy, and financing mitigation and preparedness. Clearly, these three cities learned many lessons from the Loma Prieta experience and are currently using that experience to enhance emergency management policy and planning.

Intergovernmental Coordination

Preparations for a major earthquake had been extensive throughout California for many years, but the event itself did contain a number of unexpected elements. The federal government, already overextended as a consequence of Hurricane Hugo only three weeks before, understandably had some difficulty organizing another major assistance effort in California. State and local governments faced a very challenging situation: a major earthquake affecting a large, densely populated, multijurisdictional region. Obviously, the event required considerable intergovernmental coordination. Like local emergency-response activities generally, intergovernmental coordination was not the subject of intensive study following the earthquake, but it was addressed in some reports. In one study that focused on the emergency response to both Hurricane Hugo and the Loma Prieta earthquake, Schneider characterizes intergovernmental coordination in the earthquake as reasonably effective but somewhat disorganized:

> Despite greater general preparedness, some officials still had difficulty coping with the disaster . . . local officials often were not familiar with their responsibilities or with the role of other government agencies. Some expected the federal government to do everything. More commonly, local officials tried to do things that FEMA (or some other federal agency) was supposed to do. Their actions seemed appropriate and necessary at the time, but they disrupted the functioning of the intergovernmental system.
>
> Schneider, 1990

Following Hurricane Hugo and the Loma Prieta earthquake, the U. S. General Accounting Office undertook a study of the performance of federal government agencies in disaster response and relief activities. Much of the report focuses on federal activities and responsibilities related to recovery, but emergency-response issues are also touched upon. The report assesses the response favorably, noting that "California's level of preparedness contributed to its ability to respond to the earthquake with relatively few problems" and that "A FEMA exercise that tested the catastrophic earthquake plan—two months before the earthquake—contributed greatly to a well-coordinated response" (GAO, 1991). But the report goes on to identify ways in which the response effort might have been improved, pointing out that standard operating procedures for state emergency-operations centers were inadequate or lacking; that many federal agencies did not have sufficient staff available to perform critical functions adequately; that FEMA's emphasis on war preparedness left many staff ill-prepared to provide services in disasters; and that in providing emergency assistance, the Red

Cross by its own admission was "culturally insensitive to victims, and did not have appropriate bilingual skills to serve some communities." The report argues that deficiencies in the emergency response may be due to the fact that there is no government agency (at the federal or any other level) that can monitor preparedness activities and require local jurisdictions to perform their response-related roles effectively.

CONCLUDING COMMENTS

Research on the public and organizational response to the Loma Prieta earthquake reemphasized many old lessons. Among these lessons are that disasters create an outpouring of altruism, but this massive response can in itself create coordination problems; that people behave adaptively in disaster situations, and public education can improve their chances of remaining safe; that when organizations show a real commitment to disaster preparedness, those preparedness efforts increase organizational effectiveness when disaster strikes; and that disasters invariably produce unexpected challenges for responders, calling for flexibility and the willingness to develop innovative solutions.

At the same time, the Loma Prieta earthquake also pointed to emerging problems and needs in the emergency-response area. It pointed out, for example, that as communities in the United States change and become more culturally diverse, organized efforts to provide assistance to disaster victims must also change to accommodate that diversity. It showed that when disasters exacerbate pre-existing community problems, such as housing shortages and homelessness, agencies need to have policies in place to address those problems and to be willing to innovate. The earthquake also revealed the need for better coordination among the various levels of government, particularly mechanisms to enable agencies to dispense with red tape and facilitate the deployment of resources to areas where they can do the most good. Additionally, it highlighted the fact that while some communities and states are extremely well prepared to respond in major emergencies, many others are not. With so many lives and so much property at risk, it is imperative to further explore policy mechanisms that would maintain the capacity of those states and localities May (1991) terms the "leaders" in hazard reduction, while enhancing the capacity of the "laggers."

Finally, perhaps the most important lesson of the Loma Prieta earthquake is that the investment made in mitigation and preparedness pays off. The earthquake showed that the Bay Area has made impressive progress in improving its ability to reduce damage and to cope with the problems created by earthquakes. But it would be a mistake to extrapolate from the Loma Prieta experience to larger events that the Bay Area's faults could produce. Rather than creating complacency, the earthquake should serve as a warning for what the Bay Area, other parts of California, and other earthquake-prone areas of the country can expect in future events. It should also serve as a sobering reminder to communities in California and around the country that still have not made a commitment to reducing earthquake hazards. They can expect many more severe problems in earthquakes comparable to Loma Prieta, and they may be truly devastated by larger ones.

ACKNOWLEDGMENTS

I wish to thank Linda B. Bourque for her comments on an earlier draft of this report and the staff at the National Clearinghouse for Loma Prieta Earthquake Information at the Earthquake Engineering Research Center for making their reference files available. Ron Eguchi, Tom Tobin, and Shirley Mattingly provided important documents and other information that helped me in compiling this review.

REFERENCES

Archea, J. 1990. Immediate Reactions of People in Houses. Pp. 56-64 in The Loma Prieta Earthquake: Studies of Short-Term Impacts. R. Bolin, ed. Boulder, CO: Program on Environment and Behavior, Institute of Behavioral Science, University of Colorado, Monograph #50.

Barton, A.H. 1970. Communities in Disaster: A Sociological Analysis of Collective Stress Situations. Garden City, NY: Doubleday and Company.

Bolin, R. (Ed). 1990. The Loma Prieta Earthquake: Studies of Short-Term Impacts. Boulder, CO: Program on Environment and Behavior, Institute of Behavioral Science, University of Colorado, Monograph #50.

Bolin, R. 1993. Post-Earthquake Shelter and Housing: Research Findings and Policy Implications. Pp. 107-131 in Monograph 5: Socioeconomic Impacts, K.J. Tierney and J.M. Nigg, eds. Monograph prepared for the 1993 National Earthquake Conference. Memphis, TN: Central U.S. Earthquake Consortium.

Bolin, R., and L. Stanford. 1990. Shelter and Housing Issues in Santa Cruz County. Pp. 99-108 in The Loma Prieta Earthquake: Studies of Short-Term Impacts, R. Bolin, ed. Boulder, CO: Program on Environment and Behavior, Institute of Behavioral Science, University of Colorado, Monograph #50.

Bolton, P.A. (Ed). 1993. The Loma Prieta, California Earthquake of October 17, 1989—Public Response. U.S. Geological Survey Professional Paper 1553-B. Washington, D.C.: U.S. Government Printing Office. For sale by Book and Open-File Report Sales, USGS, Denver, CO.

Bourque, L.B., C.S. Aneshensel, and J.D. Goltz. 1991. Injury and Psychological Distress Following the Whittier Narrows and Loma Prieta Earthquakes. Paper presented at the International Conference on the Impact of Natural Disasters, Los Angeles, CA, July.

Bourque, L.B., L.A. Russell, and J.D. Goltz. 1993a. Human Behavior During and Immediately After the Loma Prieta Earthquake. Pp. B3-B22 in The Loma Prieta, California, Earthquake of October 17, 1989—Public Response, P. Bolton, ed. U.S. Geological Survey Professional Paper 1553-B. Washington, D.C.: U.S. Government Printing Office. For sale by Book and Open-File Report Sales, USGS, Denver, CO.

Bourque, L.B., L.A. Russell, and J.D. Goltz. 1993b. (Forthcoming.) Experiences During and Response to the Loma Prieta Earthquake. Los Angeles, CA: School of Public Health, University of California, Los Angeles. Draft report to the Bay Area Regional Earthquake Preparedness Project.

California SSC. 1990. Planning for the Next One: Transcripts of Hearings on the Loma Prieta Earthquake of October 17, 1989. Sacramento, CA: California Seismic Safety Commission.

California SSC. 1991. Loma Prieta's Call to Action. Sacramento, CA: California Seismic Safety Commission.

City of Oakland. 1990. Loma Prieta Earthquake After Action Report. Oakland, CA: Oakland Office of Emergency Services.

City of Watsonville. 1990. Local Hazard Mitigation Plan: October 17, 1989 Earthquake. Watsonville, CA: City of Watsonville.

County of Santa Cruz. 1990. Executive Summary: Self-Evaluation of the Emergency Response to the Earthquake of October 17, 1989. County of Santa Cruz: Office of Emergency Services.

Drabek, T.E. 1986. Human System Responses to Disaster: An Inventory of Sociological Findings. New York: Springer-Verlag.

Durkin, M.E., and C.C. Thiel. 1992. Improving Measures to Reduce Earthquake Casualties. Earthquake Spectra, 8:95-113.

Durkin, M.E., C.C. Thiel, J.E. Schneider, and T. DeVriend. 1991. Injuries and Emergency Medical Response in the Loma Prieta Earthquake. Bulletin of the Seismological Society of America, 81:2143-2166.

Dynes, R.R. 1970. Organized Behavior in Disaster. Lexington, MA: Heath Lexington Books.

EERI. 1990. Loma Prieta Earthquake Reconnaissance Report. Earthquake Spectra, Supplement to Volume 6. Oakland, CA: Earthquake Engineering Research Institute.

GAO. 1991. Disaster Assistance: Federal, State, and Local Responses to Natural Disasters Need Improvement. Washington, D.C.: U.S. General Accounting Office. Report No. GAO/RCED-91-43.

Garcia, R., N. Honeycutt, and C. Van Anne. 1993. (Forthcoming.) The First Day's Response by the Oakland Fire Department to the Loma Prieta Earthquake: The Cypress Freeway Collapse. U.S. Geological Survey Professional Paper Series.

Governor's Office of Emergency Services. 1992. Proceedings: Joint Symposium on Earthquake Hazard Management in Urban Areas. Oakland: Office of Emergency Services, Bay Area Regional Earthquake Preparedness Project.

Isenberg, J. 1992. Performance of Lifelines and Emergency Response in Watsonville, CA to the Loma Prieta Earthquake. Pp. 381-390 in Proceedings of the 4th U. S.–Japan Workshop on Earthquake Disaster Prevention for Lifeline Systems, Los Angeles, Aug. 19-21, 1991. Washington, D.C.: U.S. Government Printing Office. U.S. Department of Commerce. National Institute of Standards and Technology Special Publication 840.

Jones, N.P., R. Wagner, G.S. Smith, and F. Krimgold. 1992. A Case-Control Study of the Casualties Associated with the Loma Prieta Earthquake: County of Santa Cruz. Pp. 6253-6258 in Proceedings of the Tenth World Conference on Earthquake Engineering, Madrid, Spain, July 19-24.

Jones, N.P., E.K. Noji, G.S. Smith, and R.M. Wagner. 1993. Casualty in Earthquakes. Pp. 19-68 in Monograph 5: Socioeconomic Impacts, K.J. Tierney and J.M. Nigg, eds. Monograph prepared for the 1993 National Earthquake Conference. Memphis, TN: Central U.S. Earthquake Consortium.

May, P.J. 1991. Addressing Public Risks: Federal Earthquake Policy Design. Journal of Policy Analysis and Management, 10:263-285.

Mileti, D.S., and P.W. O'Brien. 1992. Warnings During Disaster: Normalizing Communicated Risk. Social Problems 39:40-57.

Mileti, D.S., and P.W. O'Brien. 1993. Public Response to Aftershock Warnings. Pp. B31-B41 in The Loma Prieta, California, Earthquake of October 17, 1989—Public Response, P. Bolton, ed. U.S. Geological Survey Professional Paper 1553-B. Washington, D.C.: U.S. Government Printing Office. For sale by Book and Open-File Report Sales, USGS, Denver, CO.

National Center for Earthquake Engineering Research. 1992. Findings and Recommendations: Symposium on Policy Issues in the Provision of Post-Earthquake Shelter and Housing. Proceedings of conference held in Santa Cruz, CA, Jan. 26-28, 1992.

Neal, D.M. 1990a. Volunteer Organizations in Disaster: Response and Effectiveness Following the Loma Prieta Earthquake. Final Report for the Natural Hazards Research and Applications Information Center, University of Colorado. Denton, TX: Department of Sociology, University of North Texas.

Neal, D.M. 1990b. Volunteer Organization Responses to the Earthquake. Pp. 91-98 in The Loma Prieta Earthquake: Studies of Short-Term Impacts, R. Bolin, ed. Boulder, CO: Program on Environment and Behavior, Institute of Behavioral Science, University of Colorado, Monograph #50.

Oaks, S.D. 1990. The Damage Assessment Process: The Application of ATC 20. Pp. 6-15 in The Loma Prieta Earthquake: Studies of Short-Term Impacts, R. Bolin, Ed. Boulder, CO: Program on Environment and Behavior, Institute of Behavioral Science, University of Colorado, Monograph #50.

O'Brien, P.W., and D.S. Mileti. 1992. Public Response to Warnings of Loma Prieta Aftershocks. Paper presented at the annual meeting of the Earthquake Engineering Research Institute, San Francisco, CA, February.

O'Brien, P.W., and D.S. Mileti. 1993. Citizen Participation in Emergency Response. Pp. B23-B30 in The Loma Prieta, California, Earthquake of October 17, 1989—Public Response, P. Bolton, ed. U.S. Geological Survey Professional Paper 1553-B. Washington, D.C.: U.S. Government Printing Office. For sale by Book and Open-File Report Sales, USGS, Denver, CO.

Phillips, B.D. 1991. Sheltering and Housing After Loma Prieta: Some Policy Considerations. Paper prepared for the California Seismic Safety Commission. Dallas, Texas: Department of Sociology, Southern Methodist University.

Phillips, B.D., and M. Hutchins. 1991. Living in the Aftermath: Blaming Processes in the Loma Prieta Earthquake. Unpublished manuscript. Dallas, TX: Department of Sociology, Southern Methodist University.

Phillips, S.V., and J.K. Virostek. 1990. Natural Gas Disaster Planning and Recovery: The Loma Prieta Earthquake. San Francisco, CA: Pacific Gas and Electric.

Pointer, J.E., J. Michaelis, C. Saunders, J. Martchenke, C. Barton, J. Palafox, M. Kleinrock, and J.J. Calabro. 1992. The 1989 Loma Prieta Earthquake: Impact on Hospital Patient Care. Annals of Emergency Medicine 21:73-78.

Rahimi, M., and G. Azevedo. 1993. Building Content Hazards and Behavior of Mobility-Restricted Residents. Pp. B51-B62 in The Loma Prieta, California, Earthquake of October 17, 1989—Public Response, P. Bolton, ed. U.S. Geological Survey Professional Paper 1553-B. Washington, D.C.: U.S. Government Printing Office. For sale by Book and Open-File Report Sales, USGS, Denver, CO.

Rogers, E.M., M. Berndt, J. Harris, and J. Minzer. 1990. Accuracy in Mass Media Coverage. Pp. 44-54 in The Loma Prieta Earthquake: Studies of Short-Term Impacts, R. Bolin, ed. Boulder, CO: Program on Environment and Behavior, Institute of Behavioral Science, University of Colorado, Monograph #50.

Schneider, S.K. 1990. FEMA, Federalism, Hugo, and 'Frisco. Publius: The Journal of Federalism, 20:97-115.

State/Federal Hazard Mitigation Survey Team. 1990. Hazard Mitigation Opportunities for California: The State/Federal Hazard Mitigation Survey Team Report for the October 17, 1989 Loma Prieta Earthquake, California. Federal Emergency Management Agency. Report No. FEMA-845-DR-CA.

Thiel, C.C., J.E. Schneider, D. Hiatt, and M.E. Durkin. 1992. 9-1-1 EMS Process in the Loma Prieta Earthquake. Prehospital and Disaster Medicine 7:348-358.

Tierney, K.J. 1991. Emergency Medical Care Aspects of the Loma Prieta Earthquake. Paper presented at the International Symposium on Building Technology and Earthquake Hazard Mitigation, Kunming, China, March 25-29.

Tierney, K.J. 1992. Organizational Features of U.S. Lifeline Systems and Their Relevance for Disaster Management. Pp. 423-436 in Proceedings of the 4th U.S.–Japan Workshop on Earthquake Disaster Prevention for Lifeline Systems, Los Angeles, Aug. 19-21, 1991. Washington, D.C.: U.S. Government Printing Office. U.S. Department of Commerce. National Institute of Standards and Technology Special Publication 840.

Wagner, R.M., N.P. Jones, G.S. Smith, and F. Krimgold. 1993. (Forthcoming.) Study Methods and Progress Report: A Case Control Study of the Casualties Associated with the Loma Prieta Earthquake: County of Santa Cruz. U.S. Geological Survey Professional Paper Series.

DISCUSSANTS' COMMENTS: EMERGENCY PREPAREDNESS AND RESPONSE

Henry Renteria, City of Oakland

Kathleen Tierney says that the most important lesson learned from this earthquake was the importance of the investment in mitigation and preparedness—I totally agree with that. I will try to give you a view from the local government perspective.

For a manager, the biggest problems in a disaster always come from the unknown, yet the unknown is usually staring you in the face. After the Loma Prieta earthquake, local governments' biggest problem was shelter—how to deal with the pre-existing homeless problem in Oakland. In the Oakland Hills' firestorm, it was the unmanaged vegetation in the hills, the drought, and the weather conditions—pre-existing conditions that happened on a regular basis. It is necessary to look into communities, see what problems exist now, and ask how they will be exacerbated during the earthquake.

The Citizens of Oakland Respond to Emergencies program was put into place after the earthquake and was in place for the firestorm. It is patterned after the Los Angeles Fire Department program that Dr. Tierney referred to. The program goes into neighborhoods to train people to do preparedness and response activities. These trained groups (there are 5,000 people who have completed two of the three training modules and 400 who have completed all) are now a support service to police and fire. This trained group is a step above the "emergent" volunteer that needs to be managed. There will still be emergent volunteers, and so there must be a program that will manage those people—like a volunteer reception center.

To plan for emergency shelter, it is important to know the population—know what groups exist in the jurisdiction and their needs—dietary requirements, social issues, and the counseling and support needed in a major event. Include people and community-based organizations (nonprofit social service agencies and church congregation groups) into the planning process. If possible, identify an umbrella organization that these agencies work with. Prior to the earthquake, these groups were not involved, which hurt the Bay Area communities in the long-run. They must be part of the emergency planning and emergency-management organization.

The role of the media in a disaster is very important. They can be your best friend or your worst enemy. Have a media policy that spells out how the media will be dealt with before, during, and after a major event.

In the Bay Area, the emergency medical system is yet to be tested. The disasters that have occurred have not been major medical response events, but multi-casualty events. They were able to be handled well. The hospitals that could have been lost if the earthquake had lasted another 15 or 20 seconds or had been one magnitude higher survived. In Oakland, in particular, all of the hospitals are centered around one area and we anticipate a major loss of medical support in a major event. So the local government is working very closely with the county personnel responsible for emergency medical issues to establish casualty collection points and staging areas where medical and first-aid services can be offered.

I know Richard Ross is probably going to mention SP18-41 and the ability to use a standardized system of emergency management, but it is important that we all respond using the same language, terminology, and the same management techniques. Local preparedness also needs to include response and recovery.

This disaster woke up the population. Last June a special bond measure passed by 70 percent of the voting public to raise $50 million for emergency preparedness. This critical money will be used to retrofit essential facilities and to purchase search-and-rescue equipment, above-ground fire-fighting portable-hydrant systems, and other emergency-response equipment. The measure will also fund part of the Citizens of Oakland Respond to Emergencies program, which will continue to train a reserve force for the police and fire first-responders in Oakland.

Finally, the bigger the disaster, the greater the need for regional planning and preparedness. Involve organizations within a jurisdiction—local districts, businesses, churches, community-based organizations—as well as neighboring cities, counties, and states.

Richard D. Ross, Missouri Emergency Management Agency

It is very nice to be here with so many researchers and engineers—especially for those of us who are social scientists. In the Midwest and the mid-South, we social scientists have 150 years of research, without one result ever being leaked. We finally got around to the Central U.S. Earthquake Consortium a few years ago, and have had some real successes in the last decade—many of them plagiarized from our good friends in California. We are grateful for your support and assistance to our work.

In the mid-United States, the gross issues are being dealt with—rather than the very real, and oft times, discrete building, coordination, training, and response and recovery issues so well done in California. The inconsistencies of dealing with seven states and their ever-changing leadership, goals, and cooperation regarding earthquake issues are also difficult. The continuity of effort really rests with the Central U.S. Earthquake Consortium, the states, local governments, and a very responsible private sector.

Dr. Tierney's analysis spoke to systems redundancy—the Bay Area's capability to quickly reorganize traffic patterns throughout the area to allow businesses to get back on line and to allow the citizenry to return to some normalcy. In the middle states, this redundancy in roads, bridges, and public transportation systems is lacking.

Dr. Tierney's research may have more value for the middle states than for the earthquake sophisticates gathered here from California. I would expect to have this paper placed on the desks of state and local officials and to have that result in actions. Some of the issues not well addressed in the mid-United States, but critical in Dr. Tierney's paper, include: the evacuation issues; low-income and tenant issues; major social issues that the Stafford Act never envisioned (such as long-term shelter for those already homeless); the positive contributions of neighbors, friends, and family in response and clean-up; the need for better integration of local voluntary organizations into local plans and exercises; the challenges presented by the various esoteric telecommunications systems; the rationing of scarce resources; and issues of post-earthquake damage assessment of buildings and other structures.

Finally, emergency management, even in California, is local—communities succeed or fail based upon their capabilities. The state and the federal agencies frankly, fill gaps. Dr. Tierney mentioned the need to strengthen emergency-management programs throughout the country. There is, to this day, little funding for emergency management. Half of the in-state funding in the country for emergency management is found right here in California. And I'm sure that Dr. Richard Andrews would tell you that it does not meet the needs of the citizenry here.

We in the National Emergency Management Association are working on an initiative (in conjunction with the National Governors' Association) at this time for more meaningful mitigation programs. Most programs have been very modest and limited in geographic scope. There are few to no statewide or regional programs in mitigation. The concept of a multihazard mitigation and insurance program is to provide multihazard insurance on standard household policies, provide mitigation funds to all states and territories, and ensure that the insurance industry can continue to make insurance available regardless of the severity of disasters.

Richard Andrews, Governor's Office of Emergency Services

It is a pleasure to be here. In particular, I would like to thank my colleagues from Tennessee and Missouri, because they represent the kind of national dialogue about emergency management that has emerged in the last decade. This kind of discussion did not exist ten years ago. I would also like to thank the organizers of this conference for placing emergency response and disaster preparedness into a plenary session.

Dr. Tierney has done a marvelous job of summarizing a great deal of research that is of importance to those of us in emergency management. The old stereotype of emergency management—that it is a discipline of ex-military people who are only interested in doing the same thing over and over again—has long since passed from the scene. Here in California, for example, much of the research that Dr. Tierney cited was done in the Office of Emergency Services. Many emergency officials now come from the research community and pay attention to research.

Of all the research that was reported on today, the most promising and interesting are the studies of the epidemiology of injuries. It is very suggestive for the kinds of policies that need to be put in place—particularly the information about the large number of injuries that occurred in the 72-hour period after the earthquake. Emergency-management organizations need to factor this more into their planning.

The Loma Prieta earthquake showed that emergency shelter is really an emergency-response issue—it is not a recovery issue. It must be undertaken immediately and there must be a strategy. Emergency shelter and earthquake aftershocks are closely linked. In these large events, the aftershock problem can be more severe than the problem of the initial shock. It has a tremendous psychological impact upon those who have experienced the event—many of the mental health problems that have been seen (especially in children) are the consequence of the aftershocks.

The role of the media is more complicated than simply whether the media initially reports (accurately) on the disaster and whether those reports should be guides for emergency management. The presence of the media affects the nature of the disaster—and not always in the same way. In Hurricane Andrew, some of the problems that the state of Florida experienced were in part because the media underestimated the impact. In the case of the Los Angeles riots, the media in the first hours showed repeatedly that there was no police presence. This clearly exacerbated and contributed to the contagion of riot and civil disorder over the first night and into the second day. The Governor's Office of Emergency Services has done a good job working with the media on the office's preparedness efforts, but the media are not sensitive to the way they effect the nature of the disaster.

A "report card" on some of the issues that Dr. Tierney cited, and things that have been done since October 17, 1989, follows: FEMA has instituted a national Search And Rescue program, using a system that was developed here in California. There are now 25 interdisciplinary search-and-rescue teams more or less in place across the country—eight of them in California. They are trained and equipped and have responded (California units to Hurricane Iniki).

After the Whittier earthquake in 1987, the Governor's Office of Emergency Services contracted with the Applied Technology Council to develop the ATC-20 post-earthquake safety inspection guidelines. The office had completed training for engineers and others in the Bay Area about two weeks before the Loma Prieta earthquake. After the events, the office found that there was a tremendous management problem and has since hired a full-time engineer who works with them to oversee the ATC-20 process. The Office of Emergency Services signed a formal memorandum of understanding with the California Building Officials organization to also participate in this—so there is more confidence that the system is even more effective than it was in 1989.

There is a communication problem and a lack of accurate initial damage assessments. Utilizing state funds ($9 million) allocated by the legislature after the Loma Prieta earthquake, the Governor's Office of Emergency Services is putting into place a satellite-based emergency communication system (OASIS, Operational Area Satellite Information System). This system links state emergency centers with each county in the state as well as the principal scientific institutions and the FEMA Region IX headquarters. As part of this effort, the Governor's Office of Emergency Services is also developing the procedures to be used by local jurisdictions to report damages from earthquakes and other events. Four counties have served as prototypes to develop the procedures. Through the OASIS project (both the communication and software sides), the office is working to more aggressively provide assistance from the state and other allied jurisdictions when major disasters occur.

In January 1993, California enacted a bill that establishes a mandatory state-wide emergency-management system to be used by all agencies and jurisdictions in all multijurisdictional emergencies. The effort is to insure a standardized system with standard terminology, based on the Incident Command System and the Multi-Agency Coordination System (first developed by the fire services in California). The bill has teeth—if you experience a major emergency, and don't use the standardized system, you will not be eligible for public reimbursement from the state of California.

Recovery needs to be mentioned as an emergency-response issue. It is an area where a lot needs to be done. The relationship between the engineering and public policy communities must be strengthened for dealing with partially damaged buildings. It is inexcusable that there are major buildings not rebuilt, three years after the Loma Prieta earthquake. Much of this is directly related to the lack of engineering consensus about what should be done.

On the issue of earthquake insurance, the legislature passed a California Earthquake Insurance Program—which was just now repealed. There are lessons here that must be factored into a national-level program. While the idea of pre-funding disaster assistance is a very laudable goal and the idea of mitigation is one with which everyone agrees, the practicality of how to do that effectively still needs much work.

Lacy Suiter, Tennessee Emergency Management Agency

I appreciate the opportunity to be here today. We in the central United States do appreciate all the lessons we have learned from California over the years. I would like to report what has happened in a small way. Memphis has made a lot of progress in readiness to respond to the emergency itself. However, we have not done as much as we need to in retrofitting buildings, codes, and other activities.

On December 8, 1990, there was an exercise being conducted by the Center for Strategic and International Studies. It was frightening to go through that event—which included the chief of staff to the President of the United States, the Speaker of the House of Representatives, and others—because we realized that no one knew what it would take to recover from a $50–$100 billion disaster. The Stafford Act will not work for recovery from a catastrophic disaster. We also saw the lack of emphasis in the area of hazard reduction.

James Lee Witt has been nominated to be the director of FEMA by President Clinton. Witt's instructions from the President are simple: take the existing system and create an emergency-management system that is based on hazard reduction and not chest-pounding—reinvest, reengineer, and reinvent emergency management in the United States, never forgetting that the victims are also clients. We are hoping a system will be developed in this administration that will be proactive in terms of hazard reduction and not just reactive to the political demands of a given moment and place. As Tom Tobin told us this morning, luck is running out and it is necessary to get on with putting in place an adequate system.

5

Lifeline Perspective

Ronald T. Eguchi and Hope A. Seligson

ABSTRACT

This paper provides a summary of important lessons learned after the Loma Prieta earthquake. As a result of this event, a comprehensive research agenda was sponsored by the National Science Foundation (NSF) in collaboration with the other National Earthquake Hazard Reduction Program (NEHRP) agencies. In total, over 80 research projects were funded addressing seismological, geotechnical, structural, lifeline, and socioeconomic topics. This particular paper summarizes research findings resulting from studies on lifeline performance. Lifelines considered in this paper include water supply, wastewater, natural gas, oil, electric power, and communication systems. Transportation systems and port and harbor facilities are addressed in companion papers.

INTRODUCTION

After every major earthquake, there is a "window of opportunity" for researchers to advance the science of earthquake engineering. In cases where practical lessons are learned, these opportunities can result in significant changes in seismic analysis, design, and construction procedures. In order to substantiate these lessons, however, comprehensive, well-focused research is needed.

The NSF, along with the other NEHRP agencies, has been sponsoring earthquake-specific research since the 1971 San Fernando earthquake. In addition to the San Fernando earthquake, research initiatives were established after the 1985 Mexico City earthquake, the 1987 Whittier Narrows earthquake, and the 1989 Loma Prieta earthquake. Research agendas for these initiatives usually focused

135

on problems or issues uncovered as a result of these events. From this perspective, post-earthquake research has been problem focused.

The Loma Prieta earthquake offered a number of unique research opportunities. In the lifeline area, this earthquake allowed a detailed examination of seismic design procedures originally introduced as a result of the San Fernando event. In many cases, failure of lifeline systems was prevented because of these measures; in some cases, new vulnerabilities were uncovered. In general, research has been directed at explaining why certain design or construction measures work and why others do not.

Analyzing the earthquake vulnerability of our nation's lifeline systems is critical for several reasons. First, from the standpoint of replacement cost, lifelines account for approximately $4.5 trillion, or roughly 22 percent of the total built environment (Jones, 1993). Protecting these assets during natural disasters deserves special attention. Second, the recovery of cities after major natural disasters will depend in large part on the survivability of lifeline systems. As can be seen today in Florida, full recovery after Hurricane Andrew is slow, due in part to a lack of utility service. Many areas are still without electric power service. Finally, many systems are aged and ready for reconstruction or replacement. Taking advantage of the lessons learned from previous earthquakes offers an opportunity to enhance the seismic resistance of these systems. Identifying practical measures that can be applied to the seismic design, retrofit, and construction of lifeline systems is an essential first step in this overall process.

The purpose of this paper is to summarize practical lessons learned from research conducted as a result of the Loma Prieta earthquake. In particular, the emphasis is on research conducted to better understand the behavior of lifelines during earthquakes. While the lifeline area covers many different systems, several of the lessons learned apply to more than one lifeline. The applicability of these lessons to more than one lifeline will depend on whether they are connected or discrete systems. Connected systems generally include those that rely on transmission lines to convey service. Discrete lifelines may be classified as terminal or source facilities, for example, ports, harbors, and airports. In this paper, the emphasis is on connected systems, that is:

- the water supply;
- wastewater;
- natural gas;
- oil;
- electric power; and
- communication.

Even in the specialized area of lifeline earthquake engineering, it is difficult to identify all research efforts conducted as a result of an earthquake. For most government-sponsored research, the identification of ongoing efforts can usually be made by contacting the funding organizations and requesting a list of awards. This procedure was used in the preparation of this paper. The primary organization providing research money to study this earthquake was the NSF. Information on organizations and individuals who were awarded research grants by the NSF was provided by Drs. S.C. Liu and C.J. Astill. This help was greatly appreciated.

In addition to the NSF, other organizations have provided some money to study this earthquake. The third section of this paper attempts to highlight these efforts, where known. Other efforts that may have been conducted with the support of private funds are not documented here, unless otherwise noted.

Because of the wide variety of lifeline systems, it is impossible for any one individual to list and summarize the research results. For this reason, the authors contacted many individuals to solicit their input in the preparation of this paper. The authors would like to first acknowledge the support of the discussant panel: Donald B. Ballantyne, Kennedy/Jenks Consultants; Professor Thomas D. O'Rourke, Cornell University; Charles Roberts, Port of Oakland; and Steven H. Phillips, Pacific Gas and Electric.

The authors would also like to acknowledge several individuals who contributed to parts of this paper or who offered technical advice in writing certain sections. These individuals are Douglas Honegger, EQE International; Sam Swan, EQE International; Professor Anshel Schiff, Stanford University; and Alex Tang, Northern Telecom. In addition, there were many individuals who kindly furnished reports or papers, in a very timely manner. These individuals are: Professor A. H-S Ang, University of California, Irvine; Dr. Jeremy Isenberg, Weidlinger Associates; Professor James O. Jirsa, University of Texas; Professor Barclay Jones, Cornell University; Professor Jamshid Mohammadi, Illinois Institute of Technology; and Stuart Werner, Dames & Moore. To all of these individuals, the authors express a sincere thanks.

LIFELINE SYSTEM PERFORMANCE DURING THE 1989 LOMA PRIETA EARTHQUAKE

The following section presents brief summaries of lifeline system performance in the Loma Prieta earthquake. These summaries are intended to give an overview of system response, rather than list specific component damages. Table 5-1 lists the utility agencies as surveyed by the Earthquake Engineering Research Institute's reconnaissance report. For each utility, an assessment of the earthquake's impact has been made, based on reported damage, response, and recovery. In addition, available data on water and sewage pipeline failures have been included. These data were taken from a compilation by the Technical Council on Lifeline Earthquake Engineering of the American Society of Civil Engineers (ASCE/TCLEE, 1992).

TABLE 5-1   Impact of Loma Prieta Earthquake on Bay Area Utilities

Lifeline	Utility Agencies Located Within the Affected Areas	Impact of Earthquake None	Minor	Major	Number of Pipeline Repairs as Reported in ASCE/TCLEE Pipeline Database
WATER					
	Alameda County Water District	•			0
	Aldercroft Valley Water Company	•			0
	California Water Service Company	•			0
	Chemekta Park Water Company	•			0
	City of Cupertino — Water		•		4
	City of Hollister—Water			•	7
	City of San Francisco — City Water			•	70
	City of San Francisco — AWSS			•	5
	City of Santa Cruz—Water			•	78
	City of Tracy — Water		•		0
	City of Watsonville — Water			•	52
	East Bay Municipal Utility District — Water			•	133
	Idylwild Water Company		•		0
	Mountain Charlie Waterworks Inc.			•	16
	Pajaro Community Services District			•	7
	Purissima Hills Water District			•	5
	Redwood Estates Mutual Water Company			•	70
	Riva Ridge Mutual Water Company			•	1
	San Jose Water Company			•	202
	San Lorenzo Valley Water District			•	54
	Santa Clara Valley Water District			•	5
	Scotts Valley Water District			•	2
	Soquel Creek Water District			•	31
	Sunny Mesa Water District			•	1
	Sunny Slope County Water District		•		0
	Villa Del Monte Mutual Water Company			•	5
					748
SEWER/SANITATION					
	City of Hollister — Sewer			•	0
	City of Palo Alto — Regional Wastewater			•	1
	City of San Fracisco — Clean Water Program			•	5
	City of San Jose — Sewer			•	2
	City of Santa Cruz — Sewer			•	25
	City of Scotts Valley — Sewer			•	0
	City of Watsonville — Sewer			•	0
	East Bay Municipal Utility District — Wastewater			•	38
	Pajaro Community Services District		•		5
	San Mateo — Sewer (indirect)			•	0
	Santa Cruz County Sanitation District			•	11
	South Bayside System Authority — Wastewater			•	0
	Sunny Slope County Water District — Sewer		•		0
	Union Sanitary District			•	0
					87

TABLE 5-1 *Continued*

Lifeline	Utility Agencies Located Within the Affected Areas	Impact of Earthquake			Number of Pipeline Repairs as Reported in ASCE/TCLEE Pipeline Database
		None	Minor	Major	
COMMUNICATION					
Long Distance:					
AT&T			•		—
MCI		•			—
US Sprint		•			—
Local Networks:					
Continental Telephone (Gilroy)			•		—
General Telephone (Los Gatos)				•	—
Pacific Bell			•		—
POWER					
PG&E				•	—
GAS					
PG&E				•	—

Water Supply

In general, aqueduct and reservoir facilities were undamaged. No major damage to dam facilities was reported, although several minor cracks on embankments and spillways were noted. Storage tanks were damaged in Los Gatos, San Jose, Los Altos Hills, Watsonville, Sunny Mesa, San Lorenzo Valley, Scotts Valley, and the Santa Cruz Mountains. Pipeline damage was extensive in areas of ground failure, such as San Francisco's Marina District, Santa Cruz, and Watsonville. Disruption lasted as long as two weeks in the harder-hit areas.

In San Francisco, isolated damages causing the loss of contents of a 750,000 gallon tank severely impacted the city's fire-fighting capability (Figure 5-1). The flexibility of fire-suppression methods and the availability of the city's fire boat, however, minimized the impact of the tank failure.

Concern over possible contamination resulted in four cities in the epicentral area issuing "boil water" notices. The notices were in effect for one day in Los Gatos, three days in Watsonville, and seven days in Santa Cruz and San Lorenzo Valley (EERI, 1990).

Wastewater

Due to power outages and the lack of backup power at pumping plants, sewage was released into San Francisco and Monterey bays, as well as into the Pacific Ocean. These releases could have been avoided had adequate emergency power facilities been available (EERI, 1990).

FIGURE 5-1 Twenty-two structural fires were reported immediately after the Loma Prieta earthquake. The worst fire began inside a four-story apartment building in the Marina District, probably as a result of a leaking gas main.

Sewerage facilities were damaged in areas that suffered damage to water systems. This damage is less evident, however, and the lack of water service and subsequent disuse of sewer facilities delayed documentation of damage. Damage had been reported in the city of Watsonville, Scotts Valley, and Santa Cruz. Minor damage was also reported at various regional wastewater treatment facilities (Kennedy/Jenks/Chilton, 1990).

Natural Gas

Pacific Gas and Electric's (PG&E) natural gas transmission system was virtually undamaged—only two leaks were reported. Both were repaired without customer interruption. However, distribution systems in several areas were severely impacted. Over 1,000 pipeline leaks were reported system-wide, and three low-pressure systems were so heavily damaged that replacement was required. Replacement consisted primarily of insertion of plastic pipe into existing mains and services (Figure 5-2). The distribution system in the Marina District of San Francisco was replaced within one month, at a cost of $17 million. Fifty-one hundred customers were affected. Reconstruction of the Watsonville low-pressure system was complete within three weeks, affecting 166 customers. One hundred and forty customers were impacted in Los Gatos, where main restora-

FIGURE 5-2 Within the Marina District, damage to cast-iron natural gas pipe was so extensive that rather than repairing damaged pipe, new polyethylene pipe was inserted in existing mains and services.

tion was accomplished within ten days, and service restoration was complete within a month (Phillips and Virostek, 1990). Total gas system damages have been estimated at $19 million (Matsuda, 1993).

One of the most labor-intensive parts of the restoration process was the relighting of services. $7 million was incurred after the earthquake to relight pilots that had been turned off as a result of the earthquake (Matsuda, 1993). Service relights, which were accomplished within ten days, were required by 156,355 customers. The majority of relights resulted from customers unnecessarily turning off their own gas in response to hastily worded media messages, which recommended shutoff without specifying "if you smell or hear gas." At the peak of the relight effort, 1,183 servicemen were utilized. Outside utility companies contributing manpower to the relight effort included Southern California Gas, San Diego Gas and Electric, Mountain Fuel, Sierra Pacific, Northwest Natural Gas, and Washington Natural Gas (Phillips and Virostek, 1990).

Oil Refineries and Associated Facilities

Most of the region's refineries and tank farms are located along the San Francisco Bay in Alameda and Contra Costa counties. Numerous tanks at soft-soil sites were damaged, predominantly tanks that were full or nearly full. Typical damage modes included elephant's foot buckling, sometimes associated with loss of contents; damage to associated piping; and uplift of unanchored tank walls. It was reported that all leaks were contained within containment dikes and that no fires resulted (EERI, 1990).

Electric Power

Primarily as a result of direct damage to transmission substations, 1.4 million PG&E customers lost power following the Loma Prieta earthquake. Power was restored to most of San Francisco within seven hours, and all but 12,000 customers had power within two days (PG&E, 1990). In many cases, power restoration was accomplished by bypassing damaged equipment and operating with reduced levels of circuit protection. Damage to power generation and bulk transmission facilities was estimated to be $19 million, while distribution added an additional $4 million in damages (Matsuda, 1993).

Damage was severe in several 500-kV switchyards, including Moss Landing in Monterey Bay and Metcalf in the San Jose area. Seven 500-kV circuit breakers required replacement, at a cost of $700,000 each. These items are not stockpiled by PG&E, and only two came from within the PG&E system. The rest had to be obtained from various sources, including the Tennessee Valley Authority, the Los Angeles Department of Water and Power, and Southern California Edison (PG&E, 1990). Power plant damage was minor, but several plants were forced off-line by substation damage. The rapid loss of power throughout parts of the system was fortuitous in that many distribution systems were no longer energized when damaging situations, such as wrapping of lines, occurred.

Communications

The most notable impact of the earthquake on telecommunications was the monumental increase in call volume, both locally and worldwide. AT&T reported 27.8 million calls attempted to the 415 and 408 area codes the day after the earthquake. Nine and a half million of these calls were completed, more than double the normal daily volume of 3.5 million. Pacific Bell also reported heavy volume within the Bay Area—80 million calls versus the norm of 55 million. Heavy volumes led to dial tone delays, which in some cases impacted emergency communications (911) activities, and there were several days of service degradation during peak load times (EERI, 1990).

Direct damage to telecommunications facilities and equipment was limited. Most difficulties resulted from failure of backup power systems or insufficient backup power capacity.

RESEARCH ACTIVITIES INITIATED AFTER THE EARTHQUAKE

Post-earthquake research activities usually fall into one of two categories: federally sponsored research or research funded by private or semi-public organizations. In this paper, the emphasis is on research sponsored by the federal government, that is, the NSF. Other efforts, where publicly acknowledged, are also identified later in the paper.

NSF-Sponsored Research

While the risk to lifeline systems in earthquakes is generally acknowledged, it is not necessarily understood. Some research into the causes of damage and disruption of lifeline systems in earthquakes has been completed, but the amount is small relative to many other areas of earthquake hazard mitigation study. Data from the NSF Research Awards Database for earthquake hazard mitigation studies were reviewed for the years 1980–1990 (Seligson et al., 1991). Of the approximately $160 million spent, only about $18 million (11 percent) was spent on lifeline research. As noted in the original review, this figure does not incorporate monies routed by funded organizations (e.g., the National Center for Earthquake Engineering Research) into lifeline research and will, therefore, underestimate actual dollars spent on lifeline studies. However, a consistent comparison between general funding patterns and funding in the post-earthquake environment may be made by looking exclusively at NSF funding.

While recent funding levels for lifeline earthquake studies are an average of 11 percent of the total amount spent, this number may increase in the post-earthquake environment. Of the approximately $1.4 million in NSF funds awarded to study the 1987 Whittier Narrows earthquake, 24.5 percent were related to the study of lifelines (based on a title search of the NSF awards database, 1980–1990). Studies of structural performance received the greatest percentage of funds, approximately 37.7 percent of the total.

Following the Loma Prieta earthquake, the NSF funded $4.2 million in research through the Loma Prieta Initiative. These funds sponsored various engineering investigations, including studies addressing geotechnical, structural, seismological, and socioeconomic topics. The percentage breakdown by discipline is shown in Figure 5-3. Of this $4.2 million, about 15 percent was spent investigating lifeline issues, while 30.3 percent was spent on structural research.

Of the monies spent on lifeline research between 1980 and 1990, the majority has gone to investigating water and transportation system facilities. In general, water lifelines have received about 30 percent of the funding dedicated to lifeline topics. This number may be even larger, as studies of various multipurpose, multi-lifeline components are not included in this estimate. If the monies spent on pipeline and tank research were included, this number might be as high as 60 percent. In addition, 24 percent has gone to transportation studies. The

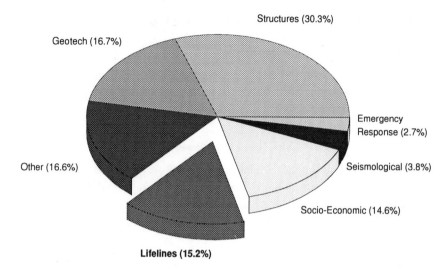

FIGURE 5-3   Breakdown of NSF Research in the Loma Prieta Initiative ($4.2 Million Total)

remainder is distributed among communications (3.3 percent), electric power (2.4 percent), natural gas (1.6 percent), wastewater (0.6 percent), disaster preparedness and emergency response (3.4 percent), and other topics.

In the immediate post-earthquake environment, funding is often focused on lifeline facilities that sustained the most damage. Following the Whittier Narrows event, 35 percent of the lifeline funding went to dams, and 27.4 percent went to bridges and overpasses. The Loma Prieta Initiative allocated 39.4 percent of lifeline monies to water studies, 30.7 percent to transportation topics, 15.6 percent to power and 14.3 percent to gas. Notably absent in the Loma Prieta studies were studies of port or harbor facilities and telecommunications.

It is interesting to note the inclusion of a significant number of socioeconomic topics in the research funded by the NSF (14.6 percent). Recently, the possibility of regional events with far-reaching impacts has prompted the study of earthquake impacts on the community in terms of direct damage, utility losses, and higher order, regional economic losses. Such research has required a multidisciplinary approach and is reflected in the variety of socioeconomic topics included in the Loma Prieta Initiative. These topics include not only community response and macroeconomic effects but secondary effects, such as those on the housing and rental markets and on the earthquake insurance industry.

Other Research Efforts

While publicly funded post-earthquake research has done a great deal to improve the state of the art in earthquake engineering, additional proprietary research whose results are not readily available has also been conducted. Some of these results, however, are gradually reaching the public domain through various vehicles including:

- the Electric Power Research Institute, which has sponsored extensive research on the damage and vulnerability of facilities relevant to the electric power and nuclear industries; several workshops that serve to exchange information have been held, including a "Wide Area Disaster Preparedness Conference" (1991), which addressed such issues as performance, restoration, mitigation, and preparedness;
- the National Communications System, which co-sponsored a workshop in 1991 with the NSF entitled "Modeling the Impact of Major Earthquakes on Communications Lifelines: Research Accomplishments and Needs," and a second workshop in 1992 entitled "Assessment of State-Of-The-Art Approaches to Communication Lifeline Modeling for Earthquake Disasters; the purpose of these workshops was to "review the state of the art in modeling the effects of major earthquakes on communications lifelines and to assess the technical feasibility of developing models if none existed"; the presenters at this workshop discussed research results on various themes including seismic testing, actual performance, predictive damage and outage models, and mitigation and preparedness measures (NCS/NSF, 1991, in press); and
- conference proceedings from various professional organizations, such as the American Society of Civil Engineers/Technical Council for Lifeline Earthquake Engineering (ASCE/TCLEE) and the Earthquake Engineering Research Institute (EERI).

Lessons Learned

In this section, lessons learned as a result of research conducted after the Loma Prieta earthquake are discussed. Although many lessons have been documented, only those that have had a major impact on the analysis or design of lifeline systems are summarized. In general, lessons learned after major earthquakes fall into one of three categories:

- lessons that identify previously unknown seismic vulnerabilities;
- lessons that substantiate or contradict prior understandings of seismic vulnerability and/or design; and
- lessons that identify new variables with regard to vulnerability assessment and/or seismic design.

In all three cases, some level of research is necessary to quantify the significance of the lesson learned or finding uncovered.

The funding of post-earthquake research has been based, in part, on the amount of damage observed in the earthquake and the impact that the earthquake had on the livelihood or welfare of the region. The results of reconnaissance surveys play a significant role in establishing research priorities. As seen in the 1971 San Fernando and the 1987 Whittier Narrows earthquakes, many structures and facilities were affected by these moderately sized earthquakes. As a result, significant research efforts were initiated to refine seismic vulnerability analysis methods and improve seismic design measures. The 1992 Landers earthquake, although larger in magnitude (M7.1), caused very little damage to buildings and lifelines. The primary reason for this lack of damage was that the earthquake occurred in a sparsely populated area of southern California. Because very little damage was observed, no major research efforts were initiated after this event, other than seismological efforts.

The Loma Prieta earthquake, however, uncovered some previously unknown vulnerabilities, primarily in the area of transportation structures. In addition, sufficient data were generated to refine the understanding of the performance of a number of other structures. A wealth of data was produced for quantifying the performance of underground lifeline components, particularly in liquefaction zones. With the exception of the 1971 San Fernando earthquake, the 1989 Loma Prieta event is responsible for the largest data set on earthquake damage to underground pipelines. As will be discussed, research conducted after the Loma Prieta earthquake was invaluable in establishing the appropriate earthquake hazard mitigation strategy for pipelines located in the city of San Francisco.

The emphasis here is on practical lessons that result in a better understanding of the seismic performance of these systems. As discussed in the introduction, the lifelines covered in this paper include:

- the water supply;
- wastewater;
- natural gas;
- oil;
- electric power; and
- communication.

Water Supply and Wastewater Systems

Many of the studies funded by the NSF after the Loma Prieta earthquake focused on the performance of water and wastewater facilities. In an earlier section of this paper, an analysis of earthquake damage data collected by ASCE/TCLEE on underground water pipelines showed that a significant number of systems located within 70 miles of the epicenter suffered some level of earthquake damage (Table 5-1). Although it is difficult to estimate the total amount of dollars spent to repair these systems, it is believed that this amount is in the tens of millions.

Historically, the amount of research on water and wastewater systems, measured in dollars, is among the largest for lifelines research. This is probably due to several factors. First, these systems have been shown to be extremely vulnerable during earthquakes. Earthquake damage to these systems has been observed in virtually every modern U.S. event. Second, because most of these systems are publicly owned, there is less reluctance on the part of the utility owner or operator to reveal earthquake damage data that may be useful for research purposes. In fact, for many of the affected water utilities, subsidization of repair costs by the Federal Emergency Management Agency is based on detailed summaries of repair activities. Therefore, there are usually good data associated with the earthquake performance of these systems.

Some of the major lessons learned from research conducted on water and wastewater facilities are listed below:

The performance of underground pipelines is closely correlated with the amount of differential movement observed in liquefaction zones.

Research conducted primarily by Professor T.D. O'Rourke at Cornell University has shown that in the Marina District, where significant liquefaction ground failure occurred (Figure 5-4), the number of pipeline repairs per unit

FIGURE 5-4 Extensive ground failure was observed in the Marina District of San Francisco. In addition to major damage to sidewalks and roadways, extensive damage to many of the underground utilities was observed.

length appeared to be closely correlated with the amount of differential displacement or settlement observed in that area. It was noted that very little damage to underground pipes was observed in areas immediately outside of the Marina District. The significance of this finding is that an assessment of the seismic vulnerability of underground pipes in future earthquakes will depend, in part, on locating these potential ground failure areas and estimating the extent of permanent ground displacement.

At low strain levels, negligible slip occurs between the soil and the pipeline; as a result, ground strains and pipeline strains can be assumed to be equal.

For the past several years, Weidlinger Associates, along with other organizations, has set up an elaborate pipeline response monitoring experiment in anticipation of the Parkfield earthquake. This experiment is designed to capture needed information on the response and behavior of different kinds of underground pipelines subject to permanent ground displacement caused by surface fault rupture. Although the Loma Prieta earthquake did not produce fault rupture at the Parkfield site, it did generate low-level ground motions that caused strains in the pipe. This suggests that the assumptions used in the design of critical pipeline systems (e.g., pipe strains from ground distortion are roughly equal to or slightly less than ground strains) are reasonable, at least for small strain levels.

Computer simulations of the post-earthquake performance of the San Francisco Auxiliary Water Supply System (AWSS) correlated well with actual leakage data. Rapid loss of water supply was predicted based on the locations of actual leaks.

Extensive research on the expected performance of the AWSS system had been performed by C.R. Scawthorn (EQE International) and T.D. O'Rourke (Cornell University) prior to the Loma Prieta earthquake. The purpose of this research was to (1) calibrate system models developed from data provided by the San Francisco Fire Department and (2) to simulate expected flow rates given probable leak locations caused by a hypothetical earthquake. In the 1989 Loma Prieta earthquake, one of the serious failures to the AWSS system occurred in a pipeline located in the South of Market area (Figure 5-5). As a result of this failure, water supply from the 750,000 gallon Jones Street tank was lost in 30 to 45 minutes. Computer simulations run after the earthquake verified the extreme vulnerability of water supply to leaks in this area and confirmed that total water supply in the one tank could be lost within a fraction of one hour. Important recommendations that resulted from this study include the suggestion to install automatic valves that would sense a rapid loss of water in the system and shut off the surviving water supply.

FIGURE 5-5 These specially designed cast-iron pipes were part of the AWSS in the South of Market area of San Francisco. Although there were few failures in this system, those that did occur had major impact on the fire-suppression water supply system.

Nonstructural components within water and wastewater treatment facilities are prone to damage from sloshing effects.

After the Loma Prieta earthquake, D. Ballantyne of Kennedy/Jenks Consultants performed a comprehensive survey of the earthquake performance of all water and wastewater treatment facilities in the San Francisco Bay area. The survey indicated that, for the most part, the structural systems at these facilities survived with little or no damage. In contrast, many nonstructural components that are critical to the effective operation of these facilities were severely impacted. It was found that sloshing effects, which are typically considered in the analysis and design of water storage tanks, can cause significant damage to other treatment plant facilities. It was concluded that much of the observed damage could be eliminated if the same principles used in the design of storage tanks were also employed for these other components.

Natural Gas Systems

Research on natural gas system performance was sponsored primarily by the following organizations: PG&E, Southern California Gas Company (SCG), and the NSF. A summary report of PG&E's response to the earthquake was prepared by PG&E (Phillips and Virostek, 1990). A detailed examination of gas system repairs in the city of San Francisco was funded by SCG and reported in a paper by Douglas G. Honegger (Honegger, 1991). An NSF-sponsored investigation of the causes of fires following the Loma Prieta earthquake was performed at the Illinois Institute of Technology (Mohammadi et al., 1992). Another NSF study, aimed at correlating gas system performance with other earthquake effects, is currently being prepared by the authors of this paper.

Key lessons learned from these studies are

Natural gas transmission pipelines and distribution mains demonstrated a high degree of ruggedness when large permanent ground deformations were absent, similar to performance in past earthquakes.

Experience during the Loma Prieta earthquake confirmed the ruggedness of buried, welded steel pipeline systems located in competent soils. The PG&E high-pressure transmission system suffered only two cracked welds in a 12-inch-diameter, 1930s vintage pipeline, which were repaired without interruption of service. Of the 25 distribution main repairs made in San Francisco, 23 were to older cast-iron pipe, and 20 were in areas known to have experienced permanent ground deformation (Phillips and Virostek, 1990; Honegger, 1991).

Damage to gas distribution lines in the city of San Francisco was largely limited to areas that experienced permanent ground deformation resulting from liquefaction, slope failure, and settlement of alluvial fill.

Examination of patterns of repair to the gas system in the city of San Francisco highlighted the high degree of vulnerability of these systems to permanent ground deformation. Examination of repair patterns outside of the Marina District showed concentrations of damage in the general vicinity of both Market Street and Golden Gate Park (Honegger, 1991). This behavior is consistent with observations from the 1987 Whittier Narrows and 1971 San Fernando earthquakes.

A strong correlation was observed between damage to structures and repairs to gas services.

Concentrations of repairs to gas services coincided with areas of San Francisco that experienced significant structural damage. The Modified Mercalli intensity contours as drawn for this event, however, did not correlate well with repair locations. Incomplete information suggests that the correlation of gas service repairs with contours of Modified Mercalli Intensity is weak, if not nonexistent, and is an area of research deserving further attention (Honegger, 1991).

There is still much to be done to educate the public, especially residential customers, about risks related to gas leakage following earthquakes.

Despite efforts by the gas companies in California to educate the public, most of the 156,000 PG&E customer calls in the ten days following the earthquake were to relight services that were unnecessarily shut off. The ability of PG&E to facilitate the relight effort was attributable to existing arrangements with other gas utilities in California, Utah, Oregon, and Washington to provide emergency service personnel. In the study by Mohammadi et al. (1992), over 90 percent of the fires in the city of San Francisco were related to electrical wiring or equipment, stoves, candles, and unknown sources. The relatively low number of fires directly related to natural gas leakage is consistent with past surveys of fire initiation following earthquakes (e.g., URS, 1988).

Electric Power Systems

While power system performance provides some of the most visible examples of earthquake impacts on lifelines, NSF-sponsored research funding has been rather limited. Part of the reason for this is that other organizations, such as the Electric Power Research Institute, have taken the lead in sponsoring research in this area. Valuable lessons that have resulted from these efforts are as follows:

A methodology developed to estimate the reliability of electric power transmission systems has identified the need to consider causes of power outage, other than direct damage, in system design, retrofit, and emergency planning.

In response to significant power outages in areas of extensive as well as limited damage, research conducted at the University of California, Irvine (Ang et al., 1992) has been directed toward developing a methodology to estimate the likelihood of electric power transmission system failure that incorporates failures due to both direct damage and power imbalances. The methodology was tested using the PG&E system in the south San Francisco area, and the results reportedly compared favorably with actual power loss patterns. Although additional validation exercises are needed, the research serves to draw further attention to possible causes of power outage, other than direct damage to facilities.

Regardless of the level of damage, power outages in urban areas can be expected to last several days after a significant earthquake to allow for inspection of high-rise buildings for gas leaks and ignition sources.

The extended power outage in downtown San Francisco following the Loma Prieta earthquake resulted not from direct damage but from the need to perform building-by-building gas leak surveys prior to energizing the local power grid. While most of the city had power restored within a day of the earthquake, the high-rise district was without power for roughly 48 hours (EQE, 1990).

Consistently poor performance of high voltage (500-kV live tank) circuit breakers in major earthquakes has shown that while seismic strengthening schemes may be adequate for small and moderate ground motions, they may not prevent failure during significant ground motion.

Power system failures are closely linked to failure of components at high voltage substations, including live-tank circuit breakers. The vulnerability of these components is generally acknowledged, and various retrofit strategies have been tried. All seven of the 500-kV live-tank circuit breakers in the strongly shaken areas of the Loma Prieta earthquake (the Metcalf and Moss Landing substations) failed. The ceramic columns supporting the interrupter heads on these circuit breakers had been retrofitted with internal fiberglass rods to help hold the columns together under seismic loads. Although a significant number of these columns failed under the loads imposed by this earthquake, it was noted that, on a few columns at Metcalf (where the intensity of ground shaking was slightly less than that at Moss Landing) only the porcelain was shaken loose, leaving the interrupter head to be supported by the reinforcing bar. The continued integrity of the reinforcing rods implies that some level of protection is offered by this retrofit scheme (EQE, 1990).

The seismic ruggedness of 500-kV dead-tank circuit breakers was demonstrated.

Although all seven 500-kV live-tank circuit breakers at Moss Landing and Metcalf were destroyed, the 500-kV dead-tank circuit breakers at these sites were undamaged. Many California power utilities are replacing older live tank circuit breakers at critical locations with shake-table-tested dead-tank circuit breakers. The Loma Prieta earthquake confirmed that these replacement breakers can remain functional after being subjected to strong ground shaking.

Communication Systems

As stated in earlier sections of this paper, it is difficult to determine the amount of research money spent to develop seismic measures for telecommunication systems. Part of the reason for this is that the majority of this research is conducted in-house by privately owned utilities, and the results of this research are only made available through limited conferences or workshops. Nevertheless, a number of valuable lessons have been documented in recent workshops and meetings.

In the United States, the most recent attempts at summarizing relevant research on telecommunication performance during earthquakes have been the two National Communications Systems/NSF workshops. The first workshop was held in 1991 in Memphis, Tennessee; the second workshop was held last year in Seattle, Washington. In both workshops, the focus was on developing improved models for predicting telecommunication performance during earthquakes. One of the case studies used in these evaluations was the Loma Prieta earthquake.

Another source of telecommunication research has been the U.S-Japan cooperative research program. In 1992, a set of meetings were held between U.S. researchers and engineers from NTT in Japan (Tang, 1992). During these meetings, visits were made to several NTT facilities, and discussions were held on different technical topics. The goal of these meetings was to explore the possibility of a formal joint agreement between the United States and Japan in the area of lifeline earthquake engineering with an emphasis on telecommunication systems. The results of these visits are summarized in a technical report prepared by the NSF (Tang, 1992).

Based on a review of the above material, the following major lessons are documented:

Sufficient slack in fiber-optic cable can help to mitigate service disruption or failure due to large relative displacements.

One of the important lessons learned by the Japanese regarding the seismic resistance of fiber-optic cable was that sufficient slack in the cable could be used to mitigate seismic damage. Figure 5-6 shows a fiber-optic cable that was stretched when one of the upper spans of the Bay Bridge collapsed during the Loma Prieta earthquake. Although the cable was stretched, there was sufficient slack to accommodate the displacement. Only 3 of 108 fibers were damaged.

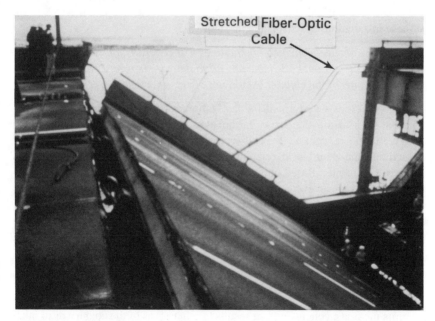

FIGURE 5-6 This photo shows a view of the collapsed upper deck of the Bay Bridge. Note in the background the fiber-optic cable that fell with the upper deck.

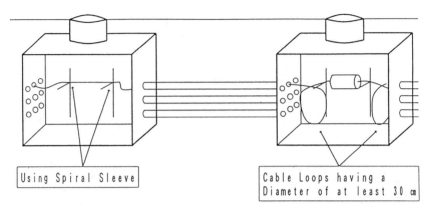

FIGURE 5-7   Earthquake countermeasures for fiber-optic cables (Yagi et al. 1992).

The Japanese are using this concept in the design of their underground manholes. One of the major concerns in the design of these facilities is that local liquefaction may uplift or displace these units. If this happens, there is concern that the cables contained within these units would either break or stretch. Therefore, in order to prevent possible failure, the Japanese have designed a number of different installations that incorporate cable slack as a design parameter. Several of these installations are displayed in Figure 5-7.

Damage to telecommunication facilities will generally affect only local communication; the national telecommunication network is robust enough that outages in one part of the country should not affect other regions.

Several studies performed for the National Communications System verified that on a nationwide scale, earthquakes should have little or no effect on calls within other regions. Several scenarios were run to determine the number of facilities that would be affected in a large earthquake on the Hayward fault in northern California. Based on fairly conservative damage criteria, it was determined that in such an event, the capacity of the network would drop to a possible low of about 67 percent in California and between 92 and 98 percent across the entire United States. These figures do not include, however, reductions in capacity caused by overload on the system.

TRANSFERABILITY OF RESULTS

One measure of the benefit of a particular research effort is the transferability of its methods, conclusions, or results to other areas of the country or to other types of facilities or systems. The transferability of earthquake research is critical, since many seismically active areas of the United States have not experienced the effects of a large earthquake. Notable areas without modern, large earthquakes include the New Madrid Seismic Zone; the Wasatch Fault Zone; the Charleston, South Carolina area; the northeastern United States; and the Pacific Northwest.

Many of the procedures available for the seismic analysis or design of lifeline systems and components are based on methods originally developed for California lifelines. In general, there are many similarities between California utility systems and systems in other parts of the United States. For example, the design and construction of major natural gas transmission pipelines, particularly interstate lines, are very similar, partly because their operation is federally regulated, and because they are designed and constructed under the same design guidelines. This is also the case for major oil pipelines. Because of these similarities, it is logical to assume that earthquakes exhibiting the same effects (such as liquefaction) would cause similar types of damage. Possible exceptions include the traveling wave effects that may be present in large midwestern earthquakes. These earthquakes may cause significantly more damage to underground pipelines or to structures sensitive to long-period effects (e.g., long-span bridges) than west coast earthquakes.

Some of the lessons learned from the Loma Prieta earthquake that are considered transferable are:

Pipeline damage models for ground failure effects (e.g., liquefaction) should be applicable in other parts of the United States.

The ground failure effects observed in northern California after the Loma Prieta earthquake can and have occurred in other parts of the United States. Extensive liquefaction ground failures were observed during the 1811 and 1812 New Madrid earthquakes. It is likely that these types of ground failures will be responsible for the majority of the damage incurred by well-designed pipelines. Therefore, pipeline damage models developed primarily from California data should be applicable to earthquakes in other parts of the United States. Perhaps the one area where there is a lack of data in pipeline response data is in characterizing the effects of wave propagation. Data from other parts of the world (e.g., the 1985 Mexico City earthquake) will be needed to quantify the effects from this phenomenon.

Potential damage to nonstructural elements in water treatment facilities could be more severe in earthquakes outside of California.

Many of the treatment facilities (storage tanks, sedimentation basins) that suffered significant damage during the Loma Prieta earthquake are considered to be long-period sensitive structures. Sloshing effects were the primary cause of damage to these facilities. In a large New Madrid event, it is expected that larger areas will be impacted by ground motions containing low-frequency energy. As a result, facilities that are sensitive to these types of frequencies would be expected to experience similar or more extensive damage. Damage may also be more significant because of the longer durations expected for midwestern earthquakes.

It is expected that after any significant U.S. earthquake, power outages in urban areas will last longer than in less developed areas, due to the need to inspect high-rise structures for gas leaks and ignition sources.

The extended power outage in downtown San Francisco following the Loma Prieta earthquake resulted not from direct damage but from the need to perform building-by-building gas leak surveys prior to energizing the local power grid. While most of the city had power restored within a day of the earthquake, the high-rise district was without power twice as long—roughly 48 hours. Similar occurrences are anticipated for other U.S. earthquakes, and the impact could be far greater, particularly for a New Madrid type of event, which has the potential to simultaneously affect numerous highly developed urban areas.

The use of cable slack in fiber-optic cable installations can reduce or eliminate potential interruption of service in any area expected to undergo significant movement or displacement.

As a practical mitigation measure, this procedure should apply to any seismic region of the world. As indicated in previous discussions, this technique has been developed by NTT engineers to mitigate the effects of liquefaction on underground manhole structures. This technique may be particularly useful in the Midwest, where extensive liquefaction is expected in large earthquakes.

FUTURE RESEARCH DIRECTIONS

As a result of lessons learned from the Loma Prieta earthquake, several encouraging trends have developed. One of the more significant efforts is the collection and documentation of pipeline failure data. Only by collecting this information will researchers be able to validate analytical models for pipeline performance or to develop empirical models that apply over a wide range of seismic hazard effects and severity levels. Therefore, continued efforts to collect and document this-data are encouraged.

Another area deserving further study is the area of ground failure assessment. It has been shown that pipeline performance is tied very closely with the types and levels of permanent ground displacement observed during an earthquake. If such a strong correlation exists, then the assessment of pipeline performance becomes a seismic hazard microzonation problem (i.e., identifying potential areas of ground failure and amounts of displacement). This further supports the continuation of efforts to map these types of hazards on a broad regional scale.

In earlier discussions of lessons learned, it was stated that system vulnerability methods were useful in identifying areas of potential outage. Ang et al. (1992) discussed the benefit of using these methods to quantify electric power system vulnerability; O'Rourke et al. (1990) used these methods to validate the vulnerability of the AWSS in the city of San Francisco and the rapid loss of water after the Loma Prieta earthquake. Although these methods have been used in areas outside of California, their application has not been widespread. One possible direction of future research is to ensure the development of appropriate regional models for system vulnerability assessment. Models that may vary from region to region include seismic hazard models (strong ground shaking, liquefaction, surface fault rupture, landslide, tsunami, and seiche) and seismic vulnerability or fragility models.

The following recommendations for collaborative research are made:

1. Stronger collaboration is needed between those researchers that characterize the severity of ground motions and ground failure effects (e.g., liquefaction) and those that model the seismic vulnerability of systems. Because lifeline systems cover large geographical areas, an assessment of seismic hazards on a broad regional scale is necessary. It is important that the appropriate seismic hazard measures are quantified and that this information is provided to the system modelers in the most useful format possible.

2. Researchers who model the performance of lifeline systems must also coordinate their studies with social scientists. It is becoming clear that to describe the full impact of these catastrophic events, it is necessary to investigate secondary and higher order effects of the event. Previous studies (ATC, 1991) have stated that the more significant losses associated with the failure of lifeline systems will come from lifeline disruption and not from repair costs. Furthermore, social disruption costs, although difficult to quantify, may also be significant. Therefore, collaboration between engineers and social scientists must be strengthened.

3. Stronger partnerships between the research community and industry must be developed. Many research efforts have benefited from such partnerships, and the results of these collaborations are more likely to lead to implementation of a study's recommendations. Currently, a federal initiative (spearheaded by the Federal Emergency Management Agency and the National Institute of Standards and Technology) to develop a plan for developing and adopting seismic design standards for public and private lifelines is underway. In order to ensure that such an effort is successful, it is essential that stronger partnerships between the research community, government agencies, and industry be developed. One of the most effective ways to initiate this partnership is to involve the end users (i.e., the lifeline operators) in government-sponsored research programs.

REFERENCES

Ang, A., H-S.J. Pires, R. Schinzinger, R. Villaverde, and I. Yoshida. 1992. *Seismic Reliability of Electric Power Transmission Systems—Applications to the 1989 Loma Prieta Earthquake.* University of California at Irvine, prepared for the National Science Foundation and the National Center for Earthquake Engineering Research.

ATC. 1991. *Seismic Vulnerability and Impact of Disruption of Lifelines in the Conterminous United States.* Applied Technology Council Report No. ATC-25, Redwood City, California.

ASCE/TCLEE. 1992. "TCLEE Pipeline Failure Database." Prepared for the National Science Foundation.

Boheim, K.B., and C.M. Kelly. 1992. "Post-Earthquake Performance of Telecommunications Networks." *Proceedings of the Fifth U.S.-Japan Workshop on Earthquake Disaster Prevention for Lifeline Systems,* Tsukuba, Japan.

EERI. 1990. *Earthquake Spectra: Loma Prieta Earthquake Reconnaissance Report,* Earthquake Engineering Research Institute, Supplement to Volume 6, May.

Electric Power Research Institute. 1991. *Proceedings: Wide-Area Disaster Preparedness Conference.* EL-7298, EPRI, Palo Alto, California.

EQE. 1990. *The October 17, 1989 Loma Prieta Earthquake: Effects on Selected Power and Industrial Facilities.* Prepared for the Electric Power Research Institute.

Honegger, D.G. 1991. "Gas System Repair Patterns in San Francisco Resulting From the 1989 Loma Prieta Earthquake." *In Proceedings of the Third U.S. Conference on Lifeline Earthquake Engineering,* Monograph No. 4, Technical Council on Lifeline Earthquake Engineering, American Society of Civil Engineers.

Isenberg, J., E. Richardson, H. Kameda, and M. Sugito. 1991. "Pipeline Response to Loma Prieta Earthquake." J. of Structural Engrg., 117(7), ASCE, New York, N.Y.

Jones, B. 1993. "New Directions in Research, Societal and Economic Studies." Presented at the 1993 Annual Meeting of the Earthquake Engineering Research Institute, Seattle, Washington.

Kennedy/Jenks/Chilton. 1990. *1989 Loma Prieta Earthquake Damage Evaluation of Water and Wastewater Treatment Facility Nonstructural Tank Elements.* Prepared for the National Science Foundation, K/J/C 896086.00.

Matsuda, E. 1993. Personal communication.

Mohammadi, J., S. Alyasin, and D.N. Bak. 1992. "Investigation of Cause and Effects of Fires Following the Loma Prieta Earthquake." Illinois Institute of Technology, Report IIT-CE-92-01, NSF Grant BCS-9003557.

NCS/NSF. 1991. *Earthquake Workshop Proceedings: Modeling the Impact of Major Earthquakes on Communication Lifelines.* Co-Sponsored by the National Communications System and the National Science Foundation.

NCS/NSF. In press. *Earthquake Workshop Proceedings: Assessment of State-of-the-Art Approaches to Communication Lifeline Modeling for Earthquake Disasters.* Co-Sponsored by the National Communications System and the National Science Foundation.

O'Rourke, T.D., H.E. Stewart, F.T. Blackburn, and T.S. Dickerman. 1990. *Geotechnical and Lifeline Aspects of the October 17, 1989 Loma Prieta Earthquake in San Francisco.* Technical Report NCEER-90-0001, NCEER, Buffalo, N.Y.

O'Rourke, T.D., J.W. Pease, and H.E. Stewart. 1992. "Lifeline Performance and Ground Deformation During the Earthquake." *The Loma Prieta, California Earthquake of October 17, 1989—Marina District.* U.S. Geological Survey Professional Paper 1551-F, Washington D.C.

PG&E. 1990. *PG&E and the Earthquake of '89,* Pacific Gas and Electric Company. San Francisco, California.

Phillips, S.H., and J.K. Virostek. 1990. *Natural Gas Disaster Planning and Recovery: The Loma Prieta Earthquake.* Pacific Gas and Electric Company, San Francisco, California.

Seligson, H.A., R.T. Eguchi, L. Lund, and C.E. Taylor. 1991. *Survey of 15 Utility Agencies Serving the Areas Affected by the 1971 San Fernando and the 1987 Whittier Narrows Earthquakes.* Prepared for the Natural Science Foundation.

Tang, A. 1992. "Technology Exchange with NTT on Seismic Protection of Telecommunication Facilities." Prepared for the National Science Foundation, ASCE/TCLEE.

URS. 1988. "Risks of Earthquake-Induced Gas Fires in Residential Housing." Report prepared for Southern California Gas Company.

Yagi, K., S. Mataki, and T. Sakurada. 1992. "Aseismic Countermeasure for Optical Fiber Cable in Liquefiable Ground." *Proceedings of the Fifth U.S.–Japan Workshop on Earthquake Disaster Prevention for Lifeline Systems*, Tsukuba Japan.

DISCUSSANTS' COMMENTS: LIFELINES

Thomas D. O'Rourke, Cornell University

It is a pleasure to be here. I have four points I would like to make, some of which will echo those made by others. Finally, I would like to give a warning about our lifeline and infrastructure systems.

Liquefaction-induced ground deformation was of key importance to the performance of the water supply in San Francisco and portions of the East Bay during the Loma Prieta earthquake. Correlations among areas of soil liquefaction and locations of buried pipeline damage show a clear pattern of system performance that depends on the severity of liquefaction and the spatial distribution of ground movement. These observations provide a practical framework for assessing the most vulnerable portions of the piping system and anticipating the effects of future earthquakes.

These spatial observations can be used to come up with some simple rules useful for planning, emergency response, and development. Since buried systems depend so much on the deformation of the ground, the fate of the ground should be looked at as being, in part, the fate of these systems. Subsurface data have been used to characterize geometry and in situ properties of loose fills in areas of San Francisco, which were subject to liquefaction and ground failure. It has been found that mapping thickness of the submerged fills, a very simple parameter, is a good indicator of the severity of damage in a given area, particularly the damage to buried lifelines. The thickness of liquefaction fill or natural sand deposit is easily used in geographical information systems, providing an excellent vehicle for assessing urban hazards, microzoning for seismic hazard reduction, and planning for optimal lifeline performance during an earthquake.

Computer simulations of damaged water supply performance during the earthquake are consistent with observations in the field and indicate that graphic modeling of hydraulic networks is sufficiently advanced for effective use in system management and emergency preparations. The computer simulations emphasize the importance of an independent power supply for isolation valves and the substantial effect that hydrant breaks have on water lost from the system.

The events of the earthquake show that flexibility provided in San Francisco by the Portable Water Supply System was of critical importance in controlling and suppressing the fire that erupted in the Marina District. The ability to operate with portable hosing and draft from a variety of water sources, including underground cisterns and fireboats, provided a valuable extra dimension in the city's emergency response.

Finally, I'd like to give a warning: beware the revenge of the infrastructure. We don't have to wait for an earthquake to have a major disaster. I hope the many valuable lessons learned after disasters can be applied to more effectively use our utility supplies and critical resources. Thank you.

Donald Ballantyne, Kennedy/Jenks Consultants

Coming from Seattle, I want to talk about what effect the Loma Prieta earthquake has had in other areas of the country. I have comments on increasing earthquake awareness, water system evaluation and design, and emergency planning.

Millions saw live TV coverage of the Loma Prieta event, which served to increase earthquake awareness. The water and sewer industry in particular had its awareness raised. The Water Pollution Control Federation was having its national conference in San Francisco that week, with many lifeline system owners from across the United States attending. They returned home to significantly influence the implementation of earthquake-mitigation programs in water and wastewater facilities. Moreover, many lifeline system owners sent teams to San Francisco to discuss the impacts of the event with their local counterparts. I believe that the closer one gets to an earthquake, the greater psychological impact it has. This can ultimately turn into the driving force to initiate earthquake-mitigation programs.

The NEHRP program identified Seattle as a target area in 1987. In 1986, there were no lifeline earthquake mitigation programs in the Seattle area. In 1993, every major water and wastewater facility has a program in place, as do many of the moderate-size utilities. This resulted from the synergy of the NEHRP program's provision to develop basic seismological data in the northwest and the focus drawn from the Loma Prieta earthquake, which demonstrated what the effects of an event could be.

Pipeline failures are concentrated in areas where liquefaction-induced permanent ground deformation occurs and can result in draining water storage tanks holding water needed for fire suppression. The old cast-iron pipe along the San Lorenzo River in Santa Cruz failed, which drained reservoirs. Service was lost to the city's two hospitals, and fire-protection capabilities were lost in those pressure zones. Luckily, fire was not a problem as there was no wind that evening. Power was not available for pumping to refill tanks for a number of days. This refocuses thinking on the hazards of liquefaction and ground deformation to vulnerable pipeline systems. System-control measures to maintain system function, and possibly dedicated fire-protection systems, are potential mitigation alternatives.

With such a large inventory of pipe in the ground and the high cost of replacement, a more reasonable approach may be isolating damaged portions of the system—either the reservoirs themselves, pipelines crossing faults, or areas that are geotechnically unstable.

Emergency planning and plan exercise is crucial. Historically, emergency planning has had a low priority. The Loma Prieta event demonstrated the need. All responsible organizations should be involved in emergency planning for their particular service. A few points particularly relevant to the water industry include provision for emergency power, pumps, chlorinators, and repair materials. Statewide or regional mutual aid programs are very worthwhile and should be improved.

Thank you.

Charles R. Roberts, Executive Director, Port of Oakland

Good morning. I am going to discuss this from the angle of implementing repair activities and some administrative issues that followed the earthquake.

There is not much information available about how to repair the damage, how to get a facility back into operation quickly, or how to repair so that a facility won't fail in the next event.

We had developed a reporting system that was activated automatically with a 5.0 event on the Hayward fault. In the event of an earthquake, supervisors informed the civil, mechanical, and electrical engineers to inspect and report back. We knew within hours where we needed to do emergency work and where we were out of business.

The next step is to have a plan to assemble a work force, establish control centers, and initiate a multichannel communications system and a pre-authorized chain of authority. The system developed by the Port of Oakland was satisfactory, except more understanding of the chain of authority needed to be emphasized.

Finally, time reporting systems, damage categories, and financial recording in accordance with and parallel to Federal Emergency Management Agency's procedures and policies must be set up ahead of time.

For implementation, we now need to take these lessons learned to explain how to repair the subsidence problems so that they will not happen again and how to repair major concrete structures standing on long piles with heavy weights.

Thank you.

Steve Phillips, PG&E

Good morning. The most important item to stress is the need to be prepared.

In 1987, PG&E developed a corporate emergency-operations-center concept designed to deal with system-wide emergencies such as an earthquake. We did go through a mock emergency exercise—using a scenario of a 7.0 earthquake on the Hayward fault.

There have been other programs—in 1984 a formal gas pipeline replacement program was developed to look at several categories of pipeline that needed to be replaced on a systematic basis (cast-iron, pre-1930 steel-distribution facilities and certain types of older transmission lines with sub-standard welds).

Although begun in 1984 and funded at $80 million, this was a 20 to 30 year program. Almost all of the leakage occurring in the PG&E system was on facilities that fell into the pipeline replacement category. Distribution systems in the Marina District, Los Gatos, and Watsonville had to be replaced. Since the Loma Prieta earthquake, a seismic risk factor (soils and proximity to fault zones) has been included into the formula for prioritizing the pipe replacement schedule.

In the mid-1980s, PG&E began to do seismic analysis of gas facilities using a pipeline scenario to see how they would fare in an event. The study had not been completed by 1989. There was no damage in those facilities—primarily because of the location and duration of the earthquake—, and PG&E was able to make the necessary modifications (which were fairly inexpensive).

On the electric side, there was a similar program also begun in the mid-1980s. The Loma Prieta earthquake validated most of the assumptions made in that study. For example, there was no substation damage at the 115-kV and below level; there was the most damage at 500-kV substations; and there was substantial but less damage at 230-kV substations. At PG&E, we also predicted we would have no damage to control room facilities. That is exactly what happened during the Loma Prieta earthquake. A prioritized replacement program was already in place for most of the targeted equipment in those facilities. Although PG&E was not completely ready when the earthquake hit, we were positioned to be able to move forward rapidly to reduce the seismic risk when the situation did occur.

Thank you.

6

Highway Bridges

James E. Roberts

INTRODUCTION

The California State Department of Transportation (CALTRANS) owns and maintains over 12,000 bridges with spans over 20 feet. There are an equal number in the city and county systems. CALTRANS maintains the condition data for all of these and some 6,000 other highway structures such as culverts (with spans under 20 feet), pumping plants, tunnels, tubes, highway patrol inspection facilities, maintenance stations, toll plazas, and other transportation-related structures. Structural details and the current condition data are maintained in the Department Bridge Maintenance files as part of the National Bridge Inventory System required by Congress and administered by the Federal Highway Administration.

These data are updated and submitted annually to the Federal Highway Administration and are the basis upon which some of the federal gas tax funds are allocated and returned to the states. The maintenance, rehabilitation, and replacement needs for bridges are prorated against the total national needs. Only this year (1993) has seismic retrofitting been accepted as an eligible item for use of federal funds, because it was assumed by most other states to be only a California problem. After much lobbying by CALTRANS, the new Federal Intermodal Surface Transportation Efficiency Act of 1992 provides for seismic retrofit to be eligible for federal bridge funds.

Immediately after the February 9, 1971, San Fernando earthquake, CALTRANS began a comprehensive upgrading of their Bridge Seismic Design Specifications and Construction Details. CALTRANS's bridge design specifications were modified to correct the identified deficiencies for application on new bridge designs. After this work was completed, the Applied Technology Council completed project ATC-6, which became the basis for a similar seismic-design specification for bridges, which was adopted by the American Association of State Highway and Transportation Officials as the national standard in 1983. Existing structures, however, proved to be a substantially more challenging problem. Research was undertaken in the United States and overseas (in New Zealand and Japan) to improve analytical techniques and to provide basic data on the strengths and deformation characteristics of lateral-load resisting systems for bridges. CALTRANS identified the vulnerable elements of existing bridges and began a statewide seismic-retrofit program for bridges to systematically reinforce the older, non-ductile bridges.

The initial phase of the CALTRANS Bridge Seismic Retrofit Program involved installation of hinge and joint restrainers to prevent deck joints from separating. Separation was the major cause of bridge collapse during the San Fernando earthquake and was judged by CALTRANS engineers and other investigators to be the highest risk to the traveling public. Included in this phase was the installation of devices to fasten the superstructure elements to the substructure in order to prevent those superstructure elements from falling off their supports. This phase was essentially completed in 1989 after approximately 1,260 bridges on the state highway system had been retrofitted at a cost of over $55 million. Funding for this program competed with other highway safety programs, which were arguably more critical in terms of statistical support. Consequently, the bridge seismic retrofit program was allocated only $4 million annually.

While the hinge and joint restrainers performed well, shear failure of columns on the I-605/I-5 separation bridge in Los Angeles during the moderate Whittier earthquake of October 1, 1987, reemphasized the inadequacies of pre-1971 column designs. Even though there was no collapse, the extensive damage resulted in plans for basic research into practical methods of retrofitting bridge columns on the existing pre-1971, non-ductile bridges. That research program had already been initiated in early 1987 at the University of California (UC), San Diego, and the Whittier earthquake merely speeded its approval and execution. Funding levels for seismic-retrofit-program implementation were increased four-fold after the Whittier earthquake to an annual level of $16 million. Even at that level, it would require some 100 years to complete the retrofitting program that is currently identified.

The Loma Prieta earthquake of October 17, 1989, again proved the reliability of hinge and joint restrainers, but the tragic loss of life at the Cypress Street Viaduct on I-880 in Oakland emphasized the necessity to immediately accelerate the column-retrofit phase of the seismic-retrofit program for bridges with a higher funding level for both research and implementation. Other structures in the earthquake-affected counties performed well, suffering the expected column damage without collapse. With the exception of a single outrigger column-cap joint confinement detail, those bridges using the post-1972 design specifications and confinement details performed well. Damage to long, multiple-level bridges showed the need to consider more carefully longitudinal resisting systems, because earthquake forces cannot be carried into abutments and approach embankments as they can on shorter bridges. After the Loma Prieta earthquake caused 44 fatalities on the state highway system, capital funding for seismic retrofitting was increased to $300 million per year. At the same time, seismic-research funding for bridges was increased from $0.5 million annually to $5.0 million annually with an initial $8.0 million allocation from the special State Emergency Earthquake Recovery legislation of November 1989, Senate Bill 36X (SB 36X). Using the special research funding provided in SB 36X, the department engaged additional research teams and facilities to assist in this massive program.

Much research has been conducted by both U.S. and foreign researchers into the causes of damage in the Loma Prieta earthquake, and much of that research is contained in the references cited in this paper. Most of the research papers can be obtained from the National Information Service for Earthquake Engineering, Earthquake Engineering Research Center at UC Berkeley. The Earthquake Engineering Research Center, located at the Richmond, California, Field Station, has been designated as the national repository for information on the Loma Prieta earthquake. There are over 175 documents on file at the respository relating to bridge aspects of the Loma Prieta earthquake. Additional research papers and project reports can be obtained from the CALTRANS Division of Structures, Sacramento, California; the Department of Applied Mechanics and Engineering Science, UC San Diego; and the National Center for Earthquake Engineering Research, State University of New York at Buffalo. The National Center for Earthquake Engineering Research has a data base search service known as QUAKLINE.

PERFORMANCE OF PRIOR RESEARCH RESULTS

Much had been learned about bridge performance in previous earthquakes (e.g., the 1971 San Fernando and 1987 Whittier earthquakes), and only bugetary constraints prevented CALTRANS from executing seismic retrofit of older bridges at a more rapid pace. It is important, however, to observe and discuss the performance of the new seismic-design criteria that had been utilized on bridges designed after 1972 and those seismic retrofit devices that had been installed prior to the Loma Prieta event.

Hinge-Joint Restraining Devices

As previously stated, the initial phase of CALTRANS' Bridge Seismic Retrofit Program involved installation of hinge and joint restrainers to prevent deck joints from separating. This was identified as the major cause of bridge collapse during the San Fernando earthquake (LeBeau et al., 1971) and in 1972 was judged by CALTRANS engineers to be the highest risk to the traveling public. Included in this phase was the installation of devices to fasten the superstructure elements to the substructure in order to resist vertical accelerations and also to prevent superstructure elements from falling off their supports. This phase was essentially completed in 1989 after approximately 1,260 bridges on the state highway system had been retrofitted at a cost of over $55 million. Research and testing of the restrainers were conducted at UC Los Angeles, by Selna and Malvar (1987). These joint restrainer systems have performed well in subsequent earthquakes, including the 1987 Whittier event (Priestley et al., 1991c), the 1989 Loma Prieta event (Mellon et al., 1993), the 1992 Cape Mendocino event (Yashinsky, 1992a), and the three most recent southern California events of 1992.

In the eight counties that were declared disaster areas after the Loma Prieta earthquake, there are approximately 350 bridges that had been retrofitted with hinge joint restrainers. There was no observed failure of any of these restrainers. CALTRANS staff engineers agree that there would have been collapse of bridge spans due to the spans falling off their supports without the installation of restrainers. Maragakis and Saiidi (1991) of the University of Nevada at Reno and Yashinsky (1990) of the CALTRANS Office of Earthquake Engineering have published papers evaluating the performance of these restrainer details. The University of Nevada at Reno was awarded a CALTRANS research project (Project R-12) to test the performance of hinge and joint-cable restrainers for bridges under dynamic loading.

Properly Confined Column Reinforcement

Most columns designed since 1971 contain a slight increase in the main-column vertical reinforcing steel and a major increase in confinement and shear reinforcing steel over the pre-1971 designs. All new columns, regardless of geometric shape, are reinforced with one or a series of spiral-wound interlocking circular cages. The typical transverse reinforcement detail now consists of #6 (.75-inch-diameter) hoops or continuous spiral at approximately 3-inch pitch over the full column height. This provides approximately eight times the confinement and shear reinforcing steel in columns than what was used in the pre-1972 non-ductile designs. All main-column reinforcing is continuous into the footings and superstructure. Splices are mostly welded or mechanical, both in the main and transverse reinforcing. Transverse-reinforcing steel is designed to produce a ductile column by confining the plastic hinge areas at the top and bottom of columns. The use of grade 60, A 706 reinforcing steel in bridges has recently been specified on a few projects on a trial basis.

In the eight counties declared disaster areas after the Loma Prieta earthquake, there are approximately 800 bridges designed after 1972 using the newly revised seismic-design criteria and confinement details. With the exception of the one outrigger beam-column joint damage on the I-980 southbound connector in Oakland, there was no documented damage to any of these 800 post-1972-designed bridges (Mellon et al., 1993).

Acceleration Response Spectra for Alluvium and Dense Foundation Materials

CALTRANS developed a series of acceleration response spectra for alluvium and average soils after the 1971 San Fernando earthquake, and these spectra were accurate for prediction of the dynamic response of those types of foundation materials. Professor Harry Bolton Seed of UC Berkeley was instrumental in the development of these design spectra. Those bridges situated on average foundation materials and designed using these spectra performed well in the Loma Prieta event.

Base-Isolated Girder Systems

Although only one bridge in the affected counties was base isolated, it did perform well during the Loma Prieta event (Mellon et al., 1993). The bridge was the Sierra Point Overhead, which was designed prior to 1972 for lateral-force requirements of only 0.06 *g*. It was subjected to lateral forces of approximately 0.18 *g* during the Loma Prieta earthquake and showed no signs of distress. It should be noted, however, that the CALTRANS design procedure is to force seismic loads into the abutments so the backwall must fail prior to the base-isolation bearings being engaged.

PROBLEMS WITH EXISTING CRITERIA, DETAILS, AND PRACTICE

A discussion of the problems encountered in highway bridge performance during the Loma Prieta earthquake will explain the need for research in the area of structural response in moderate and major earthquakes.

Older Bridges Designed for Pre-1972 Seismic Forces and Design Criteria

The major causes of bridge damage in the Loma Prieta earthquake were the criteria and details for which they were originally designed. There were over 4,000 bridges on the combined state, county, and city systems in the eight counties that were declared disaster areas after the earthquake. Only 100 of those bridges were damaged in the earthquake, and only 25 sustained what can be termed major damage, as reported in the Post Earthquake Investigation Team (PEQIT) Report (Mellon et al., 1993). Only one of the 800 bridges in the counties that had been designed after 1972, using the newer seismic forces and details, suffered damage as described in the PEQIT Report (Mellon et al., 1993).

While the Loma Prieta earthquake was, admittedly, a moderate earthquake, the bridge performance was generally what had been expected by bridge designers. Most of the research that has been commissioned since the earthquake is aimed at developing better assurance that bridges will withstand a major earthquake without collapse or major damage and that the transportation system can remain essentially functional after a major seismic event.

Seismic Performance Criteria Required

The Governor's Board of Inquiry hearings brought out the fact that there was no formal documented policy on the required seismic performance of bridges in the CALTRANS Design Specifications and Criteria (Housner, 1990). These specifications are utilized by many other public agencies, and, therefore, it is critical that a formal performance criteria be adopted.

Dynamic Response of Deep, Soft Foundations

The effect of the dynamic response of deep, soft soils in the structure foundations also proved to be a contributing factor to the collapse of the Cypress Viaduct and must be analyzed and included in future design procedures, especially for long, tall structures with relatively high periods of vibration. The effect of incoherence in the foundation response is also an important factor in the design of very long structures such as the San Francisco-Oakland Bay Bridge and the mile-long freeway viaducts. Mitchell (1992), Bolt (1991, 1992), Der Kiureghian (1991), Der Kiureghian and Neyenhofer (1992), Zafir et al. (1990), and Tamura and Shah (1991) have published research papers on this subject.

Column-Footing Interaction

Investigation of damage at the Cypress Street Viaduct in Oakland subsequent to the Loma Prieta event revealed a deficiency in many pre-1972-designed bridge footings. Some of these footings suffered joint shear failures that caused structure settlement. These footings were typically designed for vertical loads and only a 0.06 g lateral force. Subsequest investigation and research by Seible and Priestley of UC San Diego revealed a potential for failures due to lack of reinforcing steel in the top to resist lateral moments. Their conclusions, based on analysis and tests, did show a need for a top mat of reinforcing steel (Seible et al., 1992a).

Inadequate Column-Confinement Reinforcement

Other than the Cypress Viaduct failure, column damage was limited to a few critical bents on the Embarcadero Freeway Viaduct, the Terminal Separation

Structure, the Central Freeway Viaduct, and the Southern Freeway Viaduct (Route 280) (Mellon et al., 1993). Generally, those damaged bents were located in areas over deep, soft soil and bay mud. The damage on the Central Viaduct was located in a few bends on the northern end between Oak and Turk streets. This was the only damage to portions of a structure that was not constructed over deep, soft soils. These structures were closed almost immediately with the exception of that portion of the Central Viaduct south of Oak Street, where there was no sign of damage. Temporary splice beams were installed on those columns of the Central Viaduct where column hinge joints had been located in the original design. This splice was intended to keep the joint from separating in a future seismic event until a more permanent retrofit detail with new columns could be installed. A report of the analysis and recommendations was prepared for CALTRANS by Seible and Priestley (1991).

The most spectacular damage and that which was closest to collapse occurred in the vicinity of Innis Street on the Southern Freeway Viaduct, Interstate 280. The shorter of two columns supporting long outrigger bents failed in joint shear near the lower deck level. This occured at only four bent locations on the structure, however. While the damage was minimal, there was obvious concern for the integrity of these pre-1972 design, non-ductile, reinforced-concrete structures. They had all been designed in the late 1950s to early 1960s for lateral forces of 0.06 *g*, using details of the period that we now know were weak and provided insufficient confinement, especially at beam-column joints. All the damaged areas were shored up with heavy-timber falsework to reinforce them during aftershocks and possible future seismic events until permanent repairs could be made. Since the duration of the Loma Prieta earthquake was relatively short and the magnitude moderate, it was prudent to close the structures to public traffic until they could be retrofitted to current seismic standards. This damage has been reported in Cooper and Van de Pol (1991), Elsesser and Whittaker (1991), Fenves (1992), Miranda and Bertero (1991), Moehle et al. (1991), Priestley and Seible (1990), Seible and Priestley (1991), and Thewalt and Stojadinovic (1992).

Inadequate Beam-Column Joint Reinforcement

Research and analysis conducted subsequent to the Loma Prieta earthquake have shown conclusively that the lightly reinforced column-pedestal detail unique to the Cypress structure was the main cause of the total collapse. Immediately following the earthquake, the Structural Engineering Department at UC Berkeley requested that CALTRANS's Division of Structures salvage a section of the damaged Cypress Street Viaduct that had not collapsed for the purpose of conducting a series of seismic performance experiments. UC Berkeley has published reports of the experiments that were to determine the fundamental period of the complex structure under low-intensity seismic loading. Additional tests were performed to measure the actual lateral resistance of the framing system up to yield. Experiments on several proposed column-retrofit details were also conducted on this structure. Bollo et al. (1990), Miranda and Bertero (1991), Mahin and Moehle (1990; Mahin, 1991), Kay (1991), and Jones and Schroeder (1991) have published results of this analysis and research.

Inadequate Torsional Reinforcement

The China Basin Viaduct suffered bent outrigger damage at two locations in the vicinity of the Sixth Street northbound off ramp. That ramp was closed to traffic, but the mainline structure was kept open since it was a single-level structure on multiple-column bents, and no damage was observed at any other location. Emergency contracts were awarded to shore up the damaged bent and to effect a complete replacement of both the damaged bents and also the supporting columns at one bent location. This work was completed while mainline traffic was allowed to continue operating on the structure. Subsequent analysis of the outrigger performance indicated shear failure and a potential for combined torsion and shear failure, as reported by Moehle (1992; Moehle et al., 1991).

Pedestrian-Bridge Performance

During the vulnerability screening and seismic analysis of bridges subsequent to the Loma Prieta event, the CALTRANS Office of Earthquake Engineering staff has noted that the class of single-column-supported pedestrian bridges have universally required seismic-retrofit strengthening. They are generally lightweight and very narrow, offering little lateral stiffness, thus rendering them especially vulnerable to seismically induced lateral forces.

Steel-Bridge Performance

The documented failures of structural steel bridges during the earthquake were few, and they can be attributed to the date that those bridges were designed. The anchor-bolt failures on the San Mateo-Hayward Bridge, the San Francisco-Oakland Bay Bridge, and the viaducts on the east end of the Bay Bridge were major failures, but they were repaired in short order, especially the anchor bolts on the San Mateo-Hayward Bridge and on the East Bay Distribution Structure. Subsequent investigation revealed a large number of structural steel columns on the San Francisco Skyway portion of Interstate 80 and US 101 that must be strengthened to prevent collapse because of the low seismic lateral forces for which they were designed. There was no evidence of structural steel-girder failure on any of the bridges in the affected counties.

Importance Factor Applied to Critical Bridges

The Governor's Board of Inquiry, in its report of June 1, 1990, recommended that "The Department of Transportation should adopt a seismic safety policy for transportation structures that assures that transportation structures are seismically safe and that important transportation structures maintain their function after earthquakes" (Housner, 1990). During hearings conducted by the Board of Inquiry, the board discussed this issue with CALTRANS engineers. There had never been a distinction between bridges relative to their importance to the state or local community.

As a direct result of the one-month loss of the San Francisco-Oakland Bay Bridge during the Loma Prieta earthquake, it has been recommended that major transportation structures be designed to remain essentially elastic for higher seismic-force levels and longer shaking periods to reduce the damage to a nonstructural type. To accomplish this goal, a new "importance factor" was introduced into the design and retrofit performance criteria. This represents a major change in the seismic-design criteria for bridges and also represents the introduction of a subjective factor that will be based on judgement more than engineering principles.

RESEARCH IN SEISMIC RESPONSE OF BRIDGES

As a result of the problems discussed above and in response to further direction by the Governor's Board of Inquiry, which recommended that "The Department of Transportation should fund a continuing program of basic and problem-focused research on earthquake engineering issues pertinent to CALTRANS responsibilities" (Housner, 1990), and the Governor's Executive Order D-86-90, June 2, 1990, the Department of Transportation immediately accelerated its Bridge Seismic Research Program with the funding of 23 major research projects at a total cost of over $8 million. This research was "problem focused" on those areas that proved to be vulnerable during the recent earthquakes. Most of the research involved half-size model testing of bridge components and joint details. Seismic-retrofit details to strengthen existing bridges were developed and proven with this research program and their good performance in three recent earthquakes in California in April and June 1992, prove the validity of the ductile design and retrofit approach adopted by CALTRANS.

After the Loma Prieta earthquake caused 44 fatalities on the state highway system, capital funding for seismic retrofitting was increased from $4 million to $300 million per year. At the same time, seismic research funding for bridges was increased from $0.5 million annually to $5 million annually with an initial $8 million allocation from the special State Emergency Earthquake Recovery legislation of November 1989, Senate Bill 36X.

Research has been conducted, and is currently underway, at the UC San

Diego, to test and confirm the validity of several proposed design solutions for seismic retrofitting of existing single-column-bent substructure elements on bridges. Since the Loma Prieta earthquake, additional research has been and is currently being conducted at the University of California at Berkeley, San Diego, Irvine, and Davis to develop and test retrofit techniques for multiple-column bents and double-level structures, including abutment and footing details.

Development of Vulnerability-Analysis Algorithm

In order to set priorities for more than 24,000 bridges in the state for order of seismic-retrofit upgrading, CALTRANS engineering staff developed a risk-analysis procedure and adjusted it over the next three years as more information became available. Identification of bridges likely to sustain damage during an earthquake was an essential first step in the Single-Column phase of the Bridge Seismic Retrofit Program, which had begun just prior to the Loma Prieta earthquake. What can be classified as a level-one risk analysis was employed as the framework of the process that led to a consensus list of risk prioritized bridges. This risk analysis procedure was later utilized to also prioritize the multiple-column-supported bridges, but the single-column-supported bridges were deemed more vulnerable, based on experiences during the 1971 San Fernando earthquake.

A conventional risk analysis produces a probability of failure or survival. This probability is derived from a relationship between the load and resistance sides of a design equation. Not only is an approximate value for the absolute risk determined, but relative risks can be obtained by comparing the determined risks of a number of structures. Such analyses generally require vast collections of data to define statistical distributions for all, or at least the most important, elements of some form of analysis, design, or decision equations. The acquisition of this information can be costly if obtainable at all. Basically, what is done is to execute an analysis, evaluate both sides of the relevant design equation, and define and evaluate a failure or survival function. All of the calculations are carried out, taking into account the statistical distribution of every equation component designated as a variable throughout the entire procedure.

To avoid such a large, time-consuming investment in resources and to obtain results that could be applied quickly as part of the Single Column phase of the retrofit program, an alternative was recognized. What can be called a level-one risk-analysis procedure was used. The difference between a conventional and level-one risk analysis is that in a level-one analysis judgments take the place of massive data-supported statistical distributions.

The level-one risk-analysis procedure used can be summarized by the following steps:

1. *Identify major faults with high event probabilities (priority-one faults).* This step was carried out by consulting the California Division of Mines and

Geology and recent U.S. Geological Survey studies. A team of seismologists and engineers identified faults believed to be the sources of future significant seismic events. Selection criteria included location, geologic age, time of last displacement (late quarternary and younger), and length of fault (10-km minimum). Each fault recognized in step 1 was evaluated for style, length, dip, and area of faulting in order to estimate potential earthquake magnitude. Known faults were placed in one of three categories: minor (ignored for the purposes of this project), priority two (mapped and evaluated but unused for this project), or priority one (mapped, evaluated, and recognized as immediately threatening).

2. *Develop attenuation relationships at faults identified in step 1.* An average-attenuation model was developed by Mualchin and Jones (1992) of the California Division of Mines and Geology to be used throughout the state. It is the average of several published models.

3. *Define the minimum ground acceleration capable of causing severe damage to bridge structures.* The critical (i.e., damage-causing) level of ground acceleration was determined by performing nonlinear analyses on a typical, highly susceptible structure (single-column connector ramp) under varying maximum ground-acceleration loads. The lowest maximum ground acceleration that demanded the columns provide a ductility ratio of 1.3 was defined as the critical level of ground acceleration. The level of ground acceleration determined in this study was 0.5 *g*.

4. *Identify all the bridges within high risk zones defined by the attenuation model of step 2 and the critical acceleration boundary of step 3.* The shortest distance from every bridge in California to every priority-one fault was calculated. Each distance was compared with the distance from each respective level of magnitude fault to a 0.5 *g* decremented acceleration boundary. If the distance from the fault to the bridge was less than the distance from the fault to the 0.5 *g* boundary, the bridge was determined to lie in the high-risk zone and was added to the screening list for prioritization. The prioritization procedure is described below.

The CALTRANS Division of Structures has developed a computerized data base that has the coordinates of all 24,000 state, county, and city bridges stored. CALTRANS can produce a map of the entire state or any portion of the state showing the bridges, the major faults, and an overlay of the combinations. These maps can be viewed on the computer screen or printed for use by designers in screening to identify high-risk bridges. The procedure is quite simple using the computer data base to locate all highway bridges on the state system, locate all earthquake faults, then determine those structures that are in a high-risk zone.

5. *Prioritize the threatened bridges by summing weighted-bridge structural and transportation characteristic scores.* This step constitutes the process used to prioritize the bridges within the high-risk zones to establish the order of bridges to be investigated for retrofitting. It is in this step that a risk value is assigned to each bridge.

A specifically selected subset of structural and transportation characteristics of seismically threatened bridges was drawn from the CALTRANS structures computer database. Those characteristics were

- ground acceleration;
- route type (major or minor);
- average daily traffic;
- column design (single- or multiple-column bents);
- confinement details of column (relates to age);
- length of bridge;
- skew of bridge; and
- availability of detour.

Normalized pre-weight characteristic scores from 0.0 to 1.0 were assigned based on the information stored in the data base for each bridge. Scores close to 1.0 represent high-risk structural characteristics or high cost of lost transportation services. The pre-weight scores were multiplied by prioritization weights. Post-weight scores were summed to produce the assigned prioritization risk value.

Determined risk values are not to be considered exact. Due to the approximations inherent in the judgements adopted, the risks are no more accurate than the judgments themselves. The exact risk is not important. Prioritization-list qualification is determined by fault proximity and empirical attenuation data and not so much judgment. Therefore, a relatively high level of confidence is associated with the completeness of the list of threatened bridges. Relative risk is important, because it establishes the order of bridges to be investigated in detail for possible need of retrofit by designers. The risk analysis offers consistency in applying the judgments adopted to all bridges in the state.

A number of assumptions were made in the process of developing the prioritized list of seismically threatened bridges. This is typical of most engineering projects. These assumptions are based on what is believed to be the best engineering judgment available. It seems reasonable to pursue verification of these assumptions some time in the future. Two steps seem obvious: (1) monitoring the results of the design department's retrofit analyses and (2) executing a higher-level risk analysis.

Important features of this first step are the ease and cost with which it could be carried out and the data base that could be developed, highlighting bridge characteristics that are associated with structures in need of retrofit. This database will be utilized to confirm the assumptions made in the Single Column phase of the retrofit program. The same database will serve as part of the statistical support of a future conventional risk analysis as suggested in the second step. The additional accuracy inherent in a higher-order risk analysis will serve to verify previous assumptions, provide very good approximations of actual structural risk, and develop or evaluate postulated scenarios for emergency responses. It is reasonable to analyze only selected structures at this level. A manual screening process was used here, which included review of "as-built" plans by at least three engineers to identify bridges with column details that appeared to need upgrading.

After evaluating the results of the 1989 Loma Prieta earthquake, CALTRANS modified the risk analysis algorithm by adjusting the weights of the original characteristics and adding to the list. The additional characteristics are

- soil type;
- hinges (type and number);
- exposure (combination of length and average daily traffic);
- height;
- abutment type; and
- type of facility crossed.

Even though additional characteristics were added and weights were adjusted, the post-weight scores were still summed to arrive at the prioritization risk factor. The initial vulnerability priority lists for locally and state-owned bridges were produced by this technique, and retrofit projects were designed and built.

CALTRANS's Seismic Advisory Board reviewed the initial risk algorithm and suggested revisions to the procedure. During 1992, advances were made in the procedures employed by CALTRANS to prioritize bridges for seismic retrofit, and a new, more accurate algorithm was developed. The most significant improvement to the prioritization procedure is the employment of the multi-attribute decision theory. This prioritization scheme incorporates the information previously developed and utilizes the important extension to a multiplicative formulation.

This multi-attribute decision procedure assigns a priority rating to each bridge, enabling CALTRANS to more accurately decide which structures are more vulnerable to seismic activity in their current state. The prioritization rating is based on a two-level approach that separates out seismic hazard from impact and structural vulnerability characteristics. Each of these three criteria (hazard, impact, and structural vulnerability) depends on a set of attributes that have direct impact on the performance and potential losses of a bridge. Each of the criteria and attributes have assigned weights to show their relative importance. Consistent with previous work, a global utility function is developed for each attribute.

This new procedure provides a systematic framework for treating preferences and values in the prioritization decision process. The hierarchical nature of this procedure has the distinct advantage of being able to consider seismicity prior to assessing impact and structural vulnerability. If seismic hazard is low or non-existent, then the values of impact and structural vulnerability are not important, and the overall post-weight score will be low, because the latter two are added, but the sum of those two are multiplied by the hazard rating. This newly developed prioritization procedure is defensible and theoretically sound. It has been approved by the Seismic Advisory Board.

Other research efforts in conjunction with the prioritization procedure involve a sensitivity study that was performed on bridge prioritization algorithms from several states. Each procedure was reviewed in order to investigate whether or not California was neglecting any important principles. One hundred California bridges were selected as a sample population, and each bridge was independently evaluated by each of the algorithms. The 100 bridges were selected to represent California bridges with respect to the variables of the various algorithms. California, Missouri, Nevada, Washington, and Illinois have thus far participated in the sensitivity study. Gilbert (1993), Maroney and Gates (1990), and Sheng and Gilbert (1991) have reported on this research.

The final significant improvement to the prioritization procedure is the formal introduction of varying levels of seismicity. A preliminary seismic activity map for the state of California has been developed in order to incorporate seismic activity into the new prioritization procedure. In late 1992, the remaining bridges on the first vulnerability priority list were reevaluated using the new algorithm, and a significant number of bridges changed places on the priority list, but there were no obvious trends.

Figure 6-1 and Table 6-1 show the new algorithm and the weighting percentages for the various factors.

Design engineers have been assigned the final task of verifying or discrediting the prioritized bridges' need for retrofitting and then, if necessary, developing retrofit contract plans. Verification of the need for retrofit is necessary due to the possibility of prioritized bridges already being capable of withstanding the maximum credible earthquake. This will be the case when judgments made in the prioritization process prove to be conservative. Emphasis is being placed on evaluation of the total bridge during this phase, and in almost all instances a dynamic analysis is necessary to make the final judgment on whether to retrofit or not.

Establishment of Seismic Performance Criteria

Working with the new Seismic Advisory Board, CALTRANS has recently adopted bridge seismic performance criteria for design of all new bridges. These criteria are also applied to seismic retrofitting design for older bridges where they are practical. In some cases, it is more prudent to replace rather than retrofit a structure, because of age or high retrofit costs. Table 6-2 contains the newly adopted seismic performance criteria for the California State Highway System.

Development of Acceleration Response Spectra for Deep Soft Bay Muds

The damage patterns experienced during the Loma Prieta earthquake of October 17, 1989, reemphasized the importance of the influence of soil and founda-

MULTI-ATTRIBUTE DECISION PROCEDURE

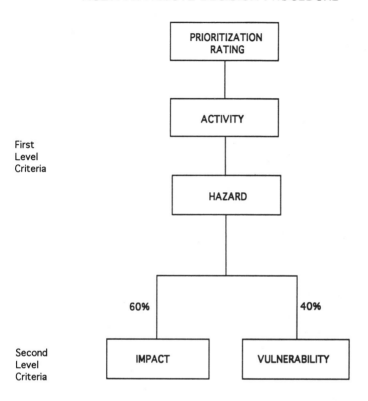

Prioritization = (Activity)(Hazard) [(0.60)(Impact) + (0.40)(Vulnerability)]
Rating

Where,

Activity = (Global Utility Function Value)

Hazard = $\sum$ (Attribute Weight)(Global Utility Function Value)

Impact = $\sum$ (Attribute Weight)(Global Utility Function Value)

Vulnerability = $\sum$ (Attribute Weight)(Global Utility Function Value)

FIGURE 6-1   Multi-attribute decision procedure—an algorithm to establish priority ratings for bridges in California.

TABLE 6-1   Multi-Attribute Decision Procedure—
Weighting Percentages for Various Criteria

HAZARD CRITERIA

Hazard Attributes	Hazard Weights
Soil Conditions	33%
Peak Rock Acceleration	38%
Seismic Duration	29%

IMPACT CRITERIA

Impact Attributes	Impact Weights
ADT on Structure	28%
ADT Under/Over Structure	12%
Detour Length	14%
Leased Air Space (Residential, Office)	15%
Leased Air Space (Parking, Storage)	7%
RTE Type on Bridge	7%
Critical Utility	10%
Facility Crossed	7%

VULNERABILITY CRITERIA

Vulnerability Attributes	Vulnerability Weights
Year Designed (Const.)	25%
Hinges (Drop Type Failure)	16.5%
Outriggers, Shared Col	22%
Bent Redundancy	16.5%
Skew	12%
Abutment Type	8%

tion response on the seismic performance of structures. Two phenomena that were highlighted were widespread soil liquefaction and the variation of recorded ground motion with local site conditions. Most structural damage occured to bridges and buildings that had been constructed over soft bay muds. The major damage at Watsonville, Santa Cruz, the South of Market district and the Marina District in San Francisco, and along the bay shores of Oakland is similar to damage patterns that occured in Mexico City in 1985. The Loma Prieta event has been called the geotechnical engineer's earthquake for that reason.

The damage to state highway bridges at Watsonville and along both the east and west shorelines of the San Francisco Bay followed that pattern as shown in the following figures. Figures 6-2 and 6-3 show the south and north sections of the Cypress Street Viaduct, respectively—the site of 42 fatalities. The south section was damaged severely but did not collapse as did the north section. Several independent investigations have concluded that a major cause of the collapse of the north section was the seismic response of the soft soils and bay mud underlying that section. As seen in Figure 6-4, the original shoreline of 100

TABLE 6-2 Seismic Performance Criteria for the Design and Evaluation of Bridges

Ground Motion at Site	Minimum Performance Level	Important Bridge Performance Level
Functional Evaluation	Immediate Service Level Repairable Damage	Immediate Service Level Minimal Damage
Safety Evaluation	Limited Service Level Significant Damage	Immediate Service Level Repairable Damage

DEFINITIONS

Immediate Service Level: Full access to normal traffic available almost immediately.

Limited Service Level: Limited access (reduced lanes, light emergency traffic) possible within days. Full service restorable within months.

Minimal Damage: Essentially elastic performance.

Repairable Damage: Damage that can be repaired with a minimum risk of losing functionality.

Significant Damage: A minimum risk of collapse, but damage that would require closure for repair.

Important Bridge (One or more of the following items present):
- Bridge required to provide secondary life safety
  (Example: access to an emergency facility)
- Time for restoration of functionality after closure creates a major economic impact
- Bridge formally designated as critical by a local emergency plan

Safety Evaluation Ground Motion (Up to two methods of defining ground motions may be used):
- *Deterministically assessed ground motions from the maximum earthquake as defined by the Division of Mines and Geology Open-File Report 92-1 (1992)*
- *Probabilistically assessed ground motions with a long return period (approximately 1,000-2,000 years)*

  For important bridges, both methods shall be given consideration, however, the probabilistic evaluation shall be reviewed by a CALTRANS approved consensus group. For all other bridges, the motions shall be based only on the deterministic evaluation. In the future, the role of the two methods for other bridges shall be reviewed by a CALTRANS approved consensus group.

Functional Evaluation Ground Motion:
  Probabilistically assessed ground motions that have a 40% probability of occurring during the useful life of the bridge. The determination of this event shall be reviewed by a CALTRANS approved consensus group. A separate Functional Evaluation is required only for Important Bridges. All other bridges are only required to meet specified design requirements to assure Minimum Functional Performance Level compliance.

FIGURE 6-2   South section of Cypress Street Viaduct.

years ago crosses the Cypress Viaduct near 14th Street. This is very near the transition between the collapsed section and the section that did not collapse. While the actual failure mode was a weak column pedestal and the low level of seismic acceleration specified in the California bridge-design specifications in use in the early 1950s, the amplified soft-soil response contributed to the complete collapse of the north portion of the viaduct. Figures 6-5 through 6-8 show the results of soft-bay-mud response on the Struve Slough Bridge near Watsonville, where the ground movement of the soft muds caused the piling to shear off below the bent caps. Figure 6-9 shows the 100-year-old shoreline on the west bay side of San Francisco Bay; most bridge damage occured to those structures that were built along the bay shore over the deep, soft muds.

FIGURE 6-3   North section of Cypress Street Viaduct.

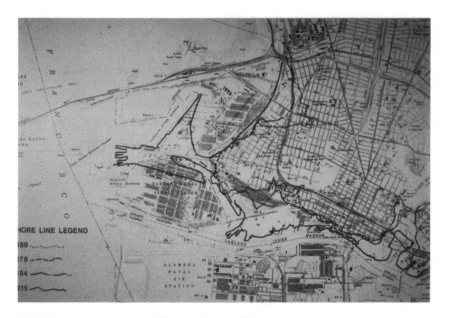

FIGURE 6-4   Present and 1880 shoreline near Cypress Structure.

FIGURE 6-5   Collapsed Struve Slough Bridge.

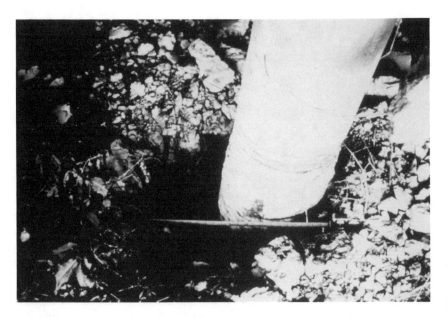

FIGURE 6-6   Evidence of large ground movement.

FIGURE 6-7   Piling sheared at girder soffit.

FIGURE 6-8   Piling penetrated six-inch deck between girder stems.

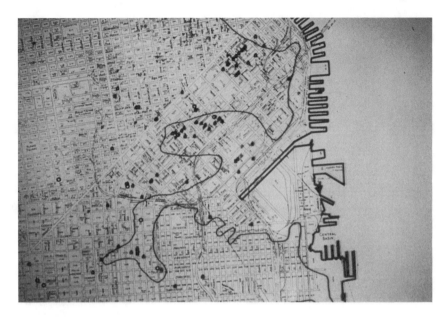

FIGURE 6-9   Present and 1880 shoreline on San Francisco side of bay.

CALTRANS developed a series of design acceleration response spectra for alluvium and normal soils after the 1971 San Fernando earthquake, but these spectra were not accurate for prediction of the dynamic response of softer soils and muds. After the Loma Prieta event, CALTRANS engaged Professors John Lysmer and Raymond B. Seed at UC Berkeley to help develop a set of similar design spectra for deep, soft, cohesionless soils and mud. Professor Seed has presented a draft report to CALTRANS, which expects to have the spectra available in 1993 (Lysmer et al., 1991). (Professor Raymond Seed's father, Professor Harry Bolton Seed, was instrumental in the development of CALTRANS's original design spectra for normal soils and alluvium.) Professor Seed et al. (1992) have concluded that the deep, soft muds amplify the bedrock ground motions by factors of two to three and that amplification of the longer-period components was especially pronounced, resulting in surface motions that are more damaging to the taller, longer period structures. Seismic ground motions have been predicted in the deep, soft soils with an analytical procedure, and the predictions have been confirmed with the actual recorded motions at four sites around the San Francisco Bay.

In addition to developing the new set of design response spectra for deep, soft soils, CALTRANS has also initiated a program to identify and map the soft soil sites in the state. While CALTRANS intends to develop a set of generic design response spectra for several representative deep, soft soil site conditions for use on the more standard and smaller bridges, it will continue to concentrate on the development of site-specific response analyses for major structures at soft soil sites, based upon the recommendation of Professor Seed and other advisors. Figure 6-10 is an example of a site-specific response spectra for a deep, soft-bay-mud site. The researchers have shown that the analysis techniques and the computer programs currently available can reliably predict the response of deep, soft soils and therefore justify site-specific analyses. Papers describing this research have been published by Jakura (1992). CALTRANS Research Project R-7 was initiated to address this area in depth.

For the major bridges crossing the bays from San Diego in the south to Antioch at the extreme northeast end of the San Francisco Bay and estuary, CALTRANS has engaged consultants to conduct site-specific complete hazard analyses using a probabilistic approach to define several levels of design earthquakes for bridge seismic-design purposes. Of the ten bridge sites in this category, these hazard studies have been completed for the San Francisco-Oakland Bay Bridge and the two bridge sites in the Carquinez Straits. Geomatrix Consultants presented their final report on this study to CALTRANS in February 1993 (Geomatrix Consultants, Inc., 1992a). Consultants have been selected for the remaining hazard studies and results will be available at various times throughout 1993.

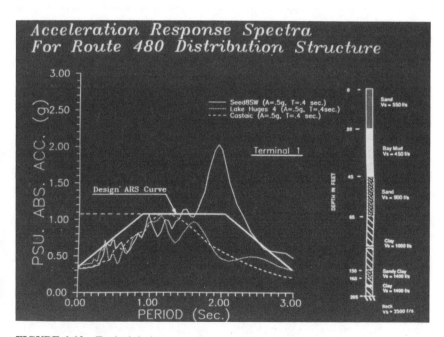

FIGURE 6-10   Typical design-response spectra for deep, soft mud.

Establishment of Procedures for Out-of-Phase
and Non-Uniform Foundation Response

Other geotechnical factors that contributed to structural damage in the Loma Prieta earthquake were non-uniform and out-of-phase responses of the foundation materials and their affect on longer structures such as the 1.5-mile-long viaducts and the 4.5-mile-long San Francisco-Oakland Bay Bridge. Geomatrix Consultants and International Civil Engineering Consultants were commissioned to conduct a coherence study for the west bay spans of the San Francisco-Oakland Bay Bridge. That study has been completed, and the report is now available (Geomatrix Consultants, Inc., 1992b). Additional research in this area has been reported by Sitar and Salgado (1992) and Eidinger and Abrahamson (1991).

Foundation Design for Liquifiable Soil Sites

While soil liquefaction was not a contributor to bridge damage, it was apparent near several major structures in the east bay and must be considered and dealt with in future seismic design for bridges. Results of studies by Lysmer et al. (1991), Idriss (1990, 1991), and other investigators clearly indicate the shortcomings of the current provisions for dealing with the influence of deep, soft foundations on structure response.

Staff seismologists, engineering geologists, and geotechnical engineers at CALTRANS have identified and mapped the deep, soft sites throughout the state, and this information has been entered into the Geographic Information System and Bridge Data Base.

Professor Geoffrey Martin at the University of Southern California has helped develop design-mitigation procedures for these liquifiable sites (Lam et al., 1991; Lam and Martin, 1991). For one site north of San Diego, CALTRANS is using 10-foot-diameter stone columns as a foundation-stiffening technique to stiffen the soil around the immediate vicinity of the bridge piers. Mitchell (1992) has published research results on ground treatment for seismic stability of bridge foundations. Research Project R-22 will also address this issue.

Pier Foundation and Abutment Soil-Structure Interaction

The response of foundation materials was a key area of needed research and the subject of several projects. In addition to the work being done by in-house staff, we have engaged researchers at both UC Davis and the University of Southern California to study the dynamic response of bridge abutments and the soil structure interaction characteristics to be used in future bridge designs. Research Project R-19 at the University of Southern Calfirnia is designed to develop improved seismic design and retrofit procedures for bridge abutments. The UC Davis study (Research Project R-20) is intended to test the soil–structure interaction and help develop reliable soil–spring constants for advanced analysis and design of this important interface under dynamic loading. Maroney et al. (1991, 1993) have published interim results of those tests. Final results of that research will be available in late 1993. Another project at UC Davis is testing the performance of bridge-foundation piles that were removed from the collapsed Struve Slough Bridge near Watsonville and retrofitted to current criteria (Research Project R-5).

CALTRANS, the city of Los Angeles, and the county of Los Angeles are supporting research at the Southern California Earthquake Research Center to determine foundation-response characteristics for the southern California area. This work will be conducted by researchers from the University of Southern Calfornia and the California Institute of Technology.

Soon after the demolition work was completed at the Cypress Street Viaduct in Oakland, CALTRANS began a series of foundation lateral-load tests. Jack Abcarius (1991a, b) describes the lateral-load tests that have been conducted on foundations at both a soft site and a hard site on the Cypress Viaduct to determine soil resistance and assist in modeling soil–spring constants for dynamic analyses. CALTRANS is also conducting a series of lateral-load tests on large-diameter piles in soft soils (CALTRANS Division of Structures, 1990) and a series of pullout capacity tests on tension piles in soft soils, where additional uplift capacity on the piles in single-column bent footings is needed to resist overturning moments. Results of these two sets of tests will be available in late spring, 1993. The test results will be used in the foundation design for replacement structures in the San Francisco area and for design of retrofitting details for the foundations of many bridges in the soft soils of the San Francisco Bay area and elsewhere in California. Mason (1993), Sweet (1993), and Wilson and Tan (1990; Wilson, 1993) have published papers on research in this area. Other research in this area is funded by Research Projects R-21 and R-27.

The University of Nevada at Reno has an ongoing research program to measure the dynamic response characteristics of a full-sized bridge on the California State Highway System near the Imperial fault east of El Centro, California. Crouse (1992), Price et al. (1992), Wilson and Tan (1990), and Gates (1993) have published several research reports on this work.

Ductile Column Design

Bridge columns designed prior to the 1971 San Fernando earthquake typically contain very little transverse reinforcement. A common detail for both circular and rectangular columns consisted of #4 (.5-inch diameter) transverse peripheral hoops at 12-inch to 18-inch centers, regardless of column size and area of main reinforcement. Also, it was common practice to extend short lengths of dowel or tails from the footing-reinforcing steel out of the footing and lap splice them with the main column-reinforcing steel cage at that point. As a consequence of these details, the ultimate curvature capable of being developed within the potential plastic region is limited by the strain at which the cover concrete starts to spall. The result is flexural failure resulting from inadequate ductility capacity or shear failure due to lack of adequate shear reinforcement. Tests conducted at UC San Diego confirm this theory (Priestley et al., 1991b). Several bridges suffered column-shear failures due to the elastic design philosophy under which they were designed prior to 1971.

Columns designed since 1971 contain a slight increase in the main-column-reinforcing steel and a major increase in confinement and shear-reinforcing steel over the pre-1971 designs. All new columns, regardless of geometric shape, are reinforced with one or a series of spiral-wound interlocking circular cages. The typical transverse shear and confinement-reinforcement detail now consists of #6 (.75-inch-diameter) hoops or continuous spiral at 3-inch pitch for the full column height. This provides approximately eight times the confinement and shear-reinforcing steel in columns than what was used in the pre-1971 non-ductile designs. All main-column reinforcing is continuous into the footings and superstructure. Splices are mostly welded or mechanical, both in the main and transverse reinforcing. Transverse-reinforcing steel is designed to produce a ductile column by confining the plastic hinge areas at the top and bottom of columns. The use of grade 60, A 706 reinforcing steel in bridges has recently been specified on a few projects on a trial basis and will probably become common practice by the end of 1993.

Research has been conducted to confirm the design criteria that were adopted after the 1971 San Fernando earthquake. Priestley and Seible (1990; Seible et al., 1992b) and Moehle and Aschheim (1992) have published research reports on this subject.

Retrofit-Strengthening Procedures for Existing Non-Ductile Columns

The largest number of large-scale tests have been conducted to confirm the calculated ductile performance of older, non-ductile bridge columns that have been strengthened by application of structural steel plate, pre-stressed strand, and fiberglass-composite jackets to provide the confinement necessary to ensure ductile performance. Research Projects R-1, R-4, R-8, and R-14 were commissioned to produce solutions to this problem. Since the spring of 1987, the researchers at UC San Diego have completed 30 sets of tests on bridge-column models. Priestley, Seible, and others at UC San Diego have published numerous research reports on this work.

UC San Diego was chosen for the testing because they had a new structural engineering laboratory with a five-story strong wall and other necessary equipment to conduct the tests. More important, the principal investigator, Dr. M.J. Nigel Priestley, had recently moved from Canterbury University in Christchurch, New Zealand, where he had been testing smaller-scale models of bridge columns using a steel-jacket concept for exterior confinement of older, non-ductile columns. With that combination of research background and the testing facility, CALTRANS awarded contracts to UC San Diego early in 1987 and had completed several tests by the time the Whittier earthquake of October 1987 again showed the weakness of pre-1971 column designs. The test series was accelerated after the Whittier earthquake, and by the time of the October 1989 Loma Prieta disaster CALTRANS had enough test results from Dr. Priestley's work that plans were on the shelf ready to be advertised for the first column-retrofit projects. Those first projects were advertised within two months of the Loma Prieta event, and work was started in early 1990 and completed by the end of that year. Figure 6-11 is a completed column-jacket-retrofit detail, the end result of this research.

Work at UC San Diego consisted of half-scale model testing of the various single-column bent retrofit techniques. Theoretical calculations and research work previously conducted in New Zealand by Dr. Priestley showed that enclos-

FIGURE 6-11   Steel jacket retrofit detail on older concrete column.

ing the columns in steel casings could significantly increase their shear strength and ductility by providing the additional confinement at the plastic hinge areas. A series of tests have been completed on round columns with outstanding results. Based on this work, a series of contracts for bridge-column retrofit were advertised in January 1990, and work is currently underway in the Los Angeles basin and the San Francisco Bay area. A second series of tests was begun in February 1990 on rectangular single-column bents, and the results were available within months. Both series of tests include models of the prototype columns with the pre-1971 reinforcing details without retrofitting, retrofitted columns using the steel-shell confinement, and a post-damage-retrofitted column using the steel shell to determine whether a non-retrofitted damaged column can be salvaged after an earthquake. Typical displacement ductility factors on retrofitted undamaged columns are 6 to 8. On the post-damage-retrofitted column, a ductility factor of 2 was achieved. Even though displacement ductility factors of 6 to 8 have been common in these first tests, the analysis procedure is based on moment ductility demand no greater than 4. Priestley et al. (1991a, b) have published several reports on this research.

Beam-Column Joint Confinement

Major advances have been made in the area of beam-column joint confinement that are based on the results of research at both UC Berkeley and UC San Diego. The performance and design criteria and structural details developed for the I-480 Terminal Separation Interchange and the I-880 replacement structures reflect the results of this research and were reported by Cooper (1992). Research is continuing at both institutions to refine further the design details to ensure ductile performance of these joints. Large one-half-scale models of the most critical three-way longitudinal edge beam, transverse bent cap, column-joint detail were built and tested at both UC San Diego and UC Berkeley as a major element of the joint and column testing program. Additional large-scale-model testing programs are planned for the more complex multilevel beam-column details commonly used in major highway interchange structures. Mahin and Moehle of UC Berkeley and Seible of UC San Diego have published reports of their research in this area (Mahin et al., 1992; Mahin, 1991; Moehle et al., 1991; and Seible et al., 1993).

The most elaborate and expensive tests conducted to date have been the half-scale and third-scale models of the proposed retrofit details for the double-deck viaduct structures in San Francisco. Models, using two different seismic retrofit techniques, were tested at both UC San Diego (half scale) and UC Berkeley (third scale) in order to obtain the performance characteristics of the two different details so CALTRANS's designers and consultants could make the final design and construction decisions and get the construction contracts started in 1992. Both models were nearly 50 ft long by 20 ft wide and 20 ft high. They included one-half of each span adjacent to the joint and half the column above and below the joint. Half-width of the superstructure was also modeled to recreate the actual contributing dead load and stiffness of the elements that frame into these columns. The difference in the concepts is in the connection to the edge of deck of the original structure. The UC Berkeley concept, funded in Research Project R-10, uses an integral edge-beam detail that is necessary on curved alignments, such as the Central Viaduct (US 101) in downtown San Francisco and the Alemany Interchange on US 101 in south San Francisco. This research-and-testing program was reported by Mahin et al. (1992). The UC San Diego concept is an independent edge beam that will be used on straight alignments, primarily the I-280 Southern Freeway Viaduct north of the Alemany Interchange with US 101. This research was reported by Priestley and Seible (1990).

Thirteen dynamic actuators were required to dynamically excite the UC San Diego model in all three directions. The beginning of plastic-hinge formation at the top of the lower column occurred at design ductilities, exactly where the design intended to force its location. After a ductility of 8 had been achieved, the plastic-hinge zone began to lose its cover, but the core concrete remained intact because of the heavy confinement reinforcement. Therefore, the structure should remain standing and in operational service. This level of damage is easily repaired after a seismic event without closure to public traffic. Each of these test series represents a cost of $500,000, but they are used to proof test the intended retrofitting scheme for structures totalling more than one mile in length with a replacement cost of over $200 million. Mahin and Moehle of UC Berkeley and Priestley and Seible of UC San Diego have published results of their work (Mahin et al., 1992; Mahin, 1991; Moehle, 1992; Priestley and Seible, 1990). Figure 6-12 shows the model at UC San Diego prior to testing; Figure 6-13 shows the UC Berkeley retrofit detail in the actual field installation.

Outrigger Bent Cap Performance Under Combined Bending, Shear, and Torsion

The Earthquake Engineering Research Center at the Richmond Field Station of the UC Berkeley began a series of tests on the performance of outrigger joints in 1991. They tested a model of the outrigger joint of the I-980 structure that was damaged in the Loma Prieta earthquake by using the identical prototype details. They replicated very closely the actual field damage, then tested the retrofitted joint as CALTRANS had redesigned it and confirmed the validity of the redesigned joint details. Based on these tests, column-cap joint details have been improved, column transverse reinforcement is typically continued up through the joint regions, and the joints are further confined for shear and torsion resistance. The details for these joints usually require 1 percent to 3 percent confinement reinforcing steel. Thewalt and Stojadinovic (1992) of UC Berkeley have reported on this research.

FIGURE 6-12   Half-scale model of edge beam retrofit in test stand.

FIGURE 6-13   Edge beam/joint detail on Interstate Route 280 in San Francisco.

Seismic Performance of Tall, Slender Pier Walls

UC Irvine has a series of research projects and large-scale model tests to investigate the performance of tall bridge pier walls under dynamic loading (Research Project R-13). They are also investigating the performance and shear strength of column-hinge-pin details (Research Project R-3). Haroum and Shepard of UC Irvine have reported the interim results of this research.

Performance of Column-Footing Joints Under Dynamic Loading

Both UC Berkeley and UC San Diego will be conducting tests on the column-footing joint details as other work is completed. Results of previous column-retrofit tests indicate that pre-1971 footing details need substantial improvement. These improvements have already been made in the retrofit contract plans, but the research is necessary to confirm theoretical calculations of performance and to continually improve details. Seible et al. (1992a, b) of UC San Diego recently published a report on this work.

Bond Development Length of Large-Diameter Reinforcing Steel Bars Under Dynamic Loading

Peer Review Panel concerns about the adequacy of the American Association of State Highway and Transportation Officials' (AASHTO) design code provisions for bond development length for number 14 and number 18 reinforcing steel bars, which are typically used in large bridge columns in California, prompted additional research testing. Full-size elements of the column-cap connection were built and tested at UC San Diego, and the results were reported by Seible et al. (1993) of UC San Diego in late 1992. The Earthquake Engineering Research Center of UC Berkeley recently completed the first in a series of tests on the performance of column-footing joint details. They also conducted tests of the bond development length as a function of concrete cover. Results of that test were reported by Filippou and Cheng (1992). Full-size tests of these details have been conducted because of concerns regarding scaling of such large bars. These tests will confirm the adequacy of current provisions or recommend changes to be made to the current code requirements at the AASHTO bridge engineer's meeting in May 1993.

Nonlinear Analysis Procedures for Reinforced-Concrete Members Subject to Dynamic Loads

This is an area where much additional research is needed to develop the technology to a state that it is a practical tool for analyzing complex structures. Engineers need to understand when nonlinear performance is important and when to spend the large resources needed to conduct a meaningful nonlinear analysis. Proper design and detailing cannot be effective without the correct performance characteristics. Professor Graham Powell of UC Berkeley was commissioned to provide guidelines for effective use of nonlinear structural analysis for bridge structures. This project provides guidelines and training to the CALTRANS engineers in the use of the new nonlinear analysis programs recently installed. Powell (1992a, b) has discussed this research in reports to CALTRANS. Filippou (1992) of UC Berkeley, Goudreau (1991) of Lawrence Livermore National Laboratories, and Yashinsky (1992b) of CALTRANS have also published reports on this subject.

Review and Revise Existing Seismic Design Specifications and Details

One of the research projects that was funded is the Applied Technology Council ATC-32 Project (CALTRANS Research Project R-11) to review and revise the entire CALTRANS Bridge Seismic Design Specifications. That project is in its second year, and the specific areas that need revision have been identified and work plans have been approved for the various sub-contractors to complete the project. Several Structural Engineers Association of California members and prominent university professors and seismic researchers are members of the Project Engineering Panel, which meets semi-annually to provide guidance to the project consultant and sub-contractors. Work should be complete by the end of 1993.

Base-Isolation Systems

Research on performance of base-isolation systems has been conducted at UC Berkeley and at the National Center for Earthquake Engineering Research in Buffalo, New York. Kelly et al. (1991) of UC Berkeley, Mayes (1992) of Dynamic Isolation Systems, and Constantinou of the National Center for Earthquake Engineering Research have reported on research in this area.

Tension-Pile Capacity Tests

CALTRANS is currently conducting a series of tension and pullout tests for eight types of piles furnished by their manufacturers. The purpose of the research program is to determine the pullout capacity of each pile type in deep, soft muds overlying rock near Interstate 280 in San Francisco. Design values will be established for each tension-pile type and included in plans for retrofitting and new construction where overturning moments require tension piles. Mason (1993) of CALTRANS, the project manager, has reported on the test program and will publish results in the fall of 1993.

PRACTICAL LESSONS APPLICABLE
TO OTHER AREAS AND STATES

A great deal of the published research that resulted from the Loma Prieta earthquake served the useful purpose of validating many of the conclusions that the bridge-engineering community had derived empirically. There were too many researchers and too much money spent on analysis of the failures at the San Francisco-Oakland Bay Bridge and the Cypress Street Viaduct. Both these structures were designed many years prior to the development of the modern seismic design specifications for bridges. Both were designed for lateral-force requirements of 0.06 g to 0.10 g and could not be expected to withstand the seismic forces that even a moderate earthquake such as Loma Prieta produced. It is a tribute to uncalculated redundancy and better than specified material strengths that these structures and other older highway structures performed as well as they did during the Loma Prieta event. This reasonable performance of older bridges in a moderate earthquake is significant for the rest of the United States, because that knowledge can assist their engineers in designing an appropriate seismic-retrofit program for their structures. While there is a necessary concern about the "Big One" in California, especially for the performance of important structures, it must be noted that many structures that can be bypassed need not be designed or retrofitted to the highest standards. It is also important to note that there will be many moderate earthquakes that will not produce the damage associated with a maximum event. These are the earthquake levels that should be addressed first in a multi-phased retrofit-strengthening program, based on the limited resources that will be available. Cost-benefit analysis of proposed retrofit details is essential to measure and ensure the effectiveness of a program.

Nevertheless, the screening processes and prioritization procedures that have been developed in California can have immediate application to other areas. Many excellent design details have been developed, through research and model testing, to guarantee ductile performance, and these can be used elsewhere without "reinventing the wheel." The excellent work that has been accomplished in foundation response and soil-structure interaction during a seismic event represents significant improvement in the state of the art and can be directly applicable in other parts of the country.

Good Emergency-Response Plan

From the experience CALTRANS has gained over the years responding to natural disasters, and the results of its response to the Loma Prieta disaster, it is clear that a sound, well-rehearsed emergency-response plan is a must for any agency charged with public safety and mobility. Because of the emergency exercises that have been conducted by the various state and local agencies and the contracting industry, CALTRANS was able to mobilize workmen, equipment, and materials almost instantaneously to repair, shore up, and reopen vital transportation facilities. Roberts (1990) and others of CALTRANS have published papers on this subject. Every public agency should have a plan prepared and exercised so personnel know exactly where they are to report and what their responsibilities are in the event of a natural disaster. Given that an earthquake is over in seconds, there is no time to plan after the event. Response must be based on prior planning and practice exercises. The plan must include redundant routes and an assessment of important roads and structures needed for immediate recovery.

Vulnerability-Prioritization Procedure

Given the large number of bridges that most states and major counties own, there is a need for a procedure to assess the relative vulnerability of each structure and prepare a priority list for retrofit work. The most critical bridges must be reinforced and retrofitted first. There will be insufficient funds to upgrade all the bridges to the current criteria, so some trade-offs must be made between cost of retrofitting versus cost of damage repair when and if an individual bridge is damaged by an earthquake. The question that must be considered is "How much do we spend on insurance now to reduce future damage?" The great deal of research that has gone into development of a vulnerability-prioritization procedure relieves other agencies from that task. The currently used algorithms can always be improved, but there is in place today an adequate procedure that has been tested statistically with other states.

Prioritization of Phases of Retrofit Work

Retrofitting older bridges can be an expensive undertaking. CALTRANS has learned from the experiences before and after the Loma Prieta event that there are a number of measures that can significantly improve the performance of bridges for a nominal investment. It has analyzed causes of bridge failures in the large earthquakes over the past 20 years in California and has observed and analyzed the performance of bridge-retrofit details and drawn some important conclusions.

First, CALTRANS prioritized the bridges using the vulnerability prioritization procedure that is discussed above. This prioritization procedure ensures that the most vulnerable bridges are scheduled for seismic retrofitting first. This uses the limited funds where they will do the most good for the overall system.

Second, it determined that the unrestrained hinge and abutment joints were the cause of many failures in earlier earthquakes. They had been restrained with engineered joint-restrainer details prior to the Loma Prieta event, and they performed very satisfactorily. In the later Cape Mendocino and Landers earthquakes of 1992, these joint restrainers again performed well. CALTRANS had spent approximately $55 million to tie joints and superstructures together on 1,260 bridges prior to the Loma Prieta earthquake.

Third, it determined, again from analysis of actual damage in the two previous earthquakes, that the single-column supports were the next most vulnerable detail on bridges. CALTRANS began the retrofitting of single columns immediately after the Loma Prieta event and to date has spent approximately $120 million to retrofit 262 bridges that are supported with single-column bents.

Fourth, it determined that multiple-column bents will sustain major damage but, in most cases, will not collapse. There is a large amount of uncalculated ductility, even in a non-ductile design. There are multiple-load paths and multiple joints to resist lateral forces, and there is not the potential for overturning that is inherent in bridges with single-column bents. CALTRANS determined that these bridges were the third priority for retrofit and is now working on the design and construction to complete this third phase of the seismic retrofit program. It will spend another $1,300 million to complete the seismic retrofitting on the remaining 550 bridges that are supported on multiple-column bents to bring them up to the current seismic-safety criteria.

Fifth, CALTRANS knows that when this program is completed there will be many bridges left in the system that will sustain damage and even require closure in the event of a maximum credible earthquake. It will continue working its way down the vulnerability-priority list and retrofit those bridges to reduce future damage. At this stage of the program, CALTRANS will have met the seismic safety criteria, however, and future seismic-retrofit projects will again compete with other highway safety projects for limited funds.

Response of Deep, Soft Soils

One of the most important lessons learned from the Loma Prieta earthquake and the research that has followed is the importance of calculating the seismic response of soft mud foundation sites for use in design of a bridge. Geotechnical engineers and geologists have been sounding the warnings for some time, especially after the 1985 Mexico City earthquake. It required the loss of life in the Loma Prieta earthquake to spring the funding for research into foundation response in softer, cohesionless soils and bay muds. That funding has, generally, been made available, and much valuable research has been completed with additional work to follow in the development of techniques to deal with soft and liquifiable soils. Much research has been and will continue to be completed in the area of foundations and soil-structure interaction. The bridge community is now equipped with the tools to design the appropriate structures in any foundation soil condition. Other areas of the country should learn from the foundation problems experienced in the Loma Prieta event, identify similar foundation problem areas in their jurisdictions, and build on the excellent research that has been completed since the Loma Prieta event to analyze the response of their structures to these varying foundation conditions.

Ductile-Design Details

The amount of research in ductile-design details is second only to the amount of research that has been completed in foundation response. The continuation of research in this area will be aimed at reducing the conservatism that permeates ductile column and joint design in California today. Prior to the Loma Prieta earthquake, there had been little or no research into the performance of very large column-cap joints and confinement details for large joints. Satisfactory performance of those bridges that had been designed for ductile performance during the earthquake gives the bridge-design community reassurance that details can be developed to guarantee ductile column and joint performance in a major seismic event. Recent research results show that the amount of joint reinforcement that has been designed into these details since Loma Prieta can be significantly reduced. Additional research will be funded to corroborate those initial conclusions and design details will be modified accordingly.

Soil-Structure Interaction

There had been a significant amount of work in soil-structure interaction completed overseas before the Loma Prieta event, and much additional research in this area has been completed or commissioned since the earthquake. This information is available and is important in the seismic retrofitting of older bridges, because very few of these older structures were designed with any consideration for the soil-structure interaction. Significant reduction in structural-member forces can be achieved by considering the effects of the foundation resistance on the total structure response. The limited research in this area has given designers some analytical tools for modeling soil–spring constants for both piling and pile caps/footings. For new design, it is essential to consider the effects of foundation soil-structure interaction in the modeling of the system for an accurate dynamic-response analysis. Typically, the abutments can be designed to resist a large percentage of the longitudinal earthquake forces in most bridges of shorter and moderate lengths. For any bridge, a savings in column retrofitting, and reinforcing steel in new designs, can be achieved by proper consideration of the foundation-structure interaction.

Response and Retrofitting of Structural Steel Bridges

Despite the fact that structural steel is ductile, bridges that have been designed by the pre-1972 seismic specifications must be evaluated for the seismic forces expected at the site based on earthquake magnitudes as they are known today. Typically, structural steel superstructures that had been tied to their substructures with joint and hinge restrainer systems performed well as discussed in the section on steel-bridge performance earlier in this paper. However, many elevated viaducts and some smaller structures supported on structural-steel columns that were designed prior to 1972 and that will require major retrofit strengthening for them to resist modern earthquake forces over a long period of shaking have been identified. One weak link is the older rocker bearings, which will probably roll over during an earthquake. These can be replaced with modern neoprene, teflon, pot and base-isolation bearings to ensure better performance in an earthquake. Structural-steel columns can be strengthened easily to increase their toughness and ability to withstand a long period of dynamic input.

Nonlinear Analysis Procedures

Prior to the Loma Prieta event, there was little use of nonlinear analysis in the design of bridges. In order to correctly analyze bridge performance in a major earthquake of long duration, the use of nonlinear analysis techniques is mandatory. Ample research has been completed in this area to give designers the necessary tools to conduct reasonable nonlinear analyses and design structures that will perform in a ductile manner during a major earthquake with long duration. Additional work in this area will continue to improve the expertise of, and build confidence in, the bridge designers.

CITED REFERENCES

Abcarius, J.L. 1991a. Lateral Load Tests on Driven Pile Footings. Lifeline Earthquake Engineering: Proceedings of the Third U.S. Conference, Report: Technical Council on Lifeline Earthquake Engineering, Monograph 4, American Society of Civil Engineers, New York, August, 1991.

Abcarius, Jack. 1991b. Lateral Load Tests on Driven Pile Footings. Proceedings, Third Bridge Engineering Conference at Denver, Colorado, March, 1991. Report: Transportation Research Board Record 1290. TRB, Washington DC.

Bollo, M.S., S.A. Mahin, J.P. Moehle, R.M. Stephen, and X. Qi. 1990. Observations and Implications of Tests on the Cypress Street Viaduct Test Structure. Report, Earthquake Engineering Research Center, University of California at Berkeley, December.

Bolt, B.A. 1991. Development of Phased Strong-Motion Time-Histories for Structures with Multiple Supports. Earthquake Engineering: Sixth Canadian Conference, University of Toronto Press, Toronto, Canada and Buffalo, New York.

Bolt, B.A. 1992. Seismic Strong Motion Synthetics. Seismic Design and Retrofit of Bridges, Seminar Proceedings, Earthquake Engineering Research Center, University of California at Berkeley and California Department of Transportation, Sacramento, California, 1992.

CALTRANS. 1990. Field Tests of Large Diameter Drilled Shafts. Report of Lateral Load Tests, Sacramento, California, March, 1990.

Cooper, T.R., and M. Van de Pol. 1991. Bridge Damage Review—Loma Prieta Earthquake. Lifeline Earthquake Engineering: Proceedings of the Third U.S. Conference, Report: Technical Council on Lifeline Earthquake Engineering, Monograph 4, American Society of Civil Engineers, New York, August, 1991.

Cooper, T.R. 1992. Terminal Separation Design Criteria: A Case Study of Current Bridge Seismic Design and Application of Recent Seismic Design Research. Seismic Design and Retrofit of Bridges, Seminar Proceedings, Earthquake Engineering Research Center, University of California at Berkeley and California Department of Transportation, Division of Structures, Sacramento, California.

Crouse, C.B. 1992. Estimation of Foundation Stiffnesses of Meloland Road Overcrossing During Quick-Release Tests. Report: University of Nevada at Reno, Center for Civil Engineering Earthquake Research.

Der Kiureghian, A., and A. Neuenhofer. 1992. Response Spectrum Method for Incoherent Support Motions. Seismic Design and Retrofit of Bridges, Seminar Proceedings, Earthquake Engineering Research Center, University of California at Berkeley and California Department of Transportation, Division of Structures, Sacramento, California.

Der Kiureghian, A. 1991. Response Spectrum Analysis of Bridges Subjected to Differential Support Motions Based on Loma Prieta Earthquake Data. Abstract of Research in Progress.

Eidinger, J., and N. Abrahamson. 1991. Seismic Response of Long Span Bridges to Incoherent Ground Motion. Proceedings of the First Annual Seismic Research Workshop, California Department of Transportation, Sacramento, California.

Elsesser, E., and A.S. Whittaker. 1991. Earthquake Response of the San Francisco Freeway Structures During the October 17, 1989 Earthquake. Lifeline Earthquake Engineering: Proceedings of the Third U.S. Conference, Report: Technical Council on Lifeline Earthquake Engineering, Monograph 4, American Society of Civil Engineers, New York, August.

Fenves, G.L. 1992. Earthquake Analysis and Response of Multi-Span Bridges and Viaduct Structures. Proceedings: Seismic Design and Retrofit of Bridges, University of California, Berkeley, June.

Filippou, F.C. 1992. Models for Nonlinear Static and Dynamic Analysis of Concrete Freeway Structures. Seismic Design and Retrofit of Bridges, Seminar Proceedings, Earthquake Engineering Research Center, University of California at Berkeley and California Department of Transportation, Division of Structures, Sacramento, California.

Filippou, F.C., C. Cheng. 1992. Anchorage of Large Diameter Reinforcing Bars. Seismic Design and Retrofit of Bridges, Seminar Proceedings. Earthquake Engineering Research Center, University of California at Berkeley and California Department of Transportation, Division of Structures, Sacramento, California.

Gates, J. 1993. Dynamic Field Response Studies and Earthquake Instrumentation of the Meloland Road Overcrossing. In Proceedings: ASCE Structures Congress XI. Irvine, California, April.

Geomatrix Consultants, Inc. 1992a. Seismic Response Study for Proposed Benicia-Martinez Bridge, Contra Costa and Solano Counties. Draft Report to California Department of Transportation, Division of Structures.

Geomatrix Consultants, Inc. 1992b. Coherence of West Bay Spans-San Francisco-Oakland Bay Bridge. Draft Report to California Department of Transportation, Division of Structures.

Gilbert, A. 1993. Development in Seismic Prioritization of California Bridges. In Proceedings: Ninth Annual U.S./Japan Workshop on Earthquake and Wind Design of Bridges. Tsukuba Science City, Japan, May.

Goudreau, G.I. 1991. Supercomputing and Nonlinear Seismic Structural Response of Freeway Structures. Structures Congress "91", Compact Papers. American Society of Civil Engineers, New York.

Housner, G.W. 1990. Competing Against Time. C.C. Thiel, ed. Report to Governor George Deukmejian from The Governor's Board of Inquiry on the 1989 Loma Prieta Earthquake, Publications Section, Department of General Services, State of California, Sacramento, California, June.

Idriss, I.M. 1991. Earthquake Ground Motions At Soft Soil Sites. In Proceedings of Second International Conference on Recent Advances in Geotechnical Earthquake Engineering and Soil Dynamics. St. Louis, Missouri, March 11-15.

Idriss, I.M. 1990. Response of Soft Soil Sites During Earthquakes. In Proceedings of the Memorial Symposium to Honor Professor Harry Bolton Seed. Berkeley, California, May.

Jakura, K. 1992. CALTRANS Procedures for Development of Site-Specific Acceleration Response Spectra", Seismic Design and Retrofit of Bridges, Seminar Proceedings, Earthquake Engineering Research Center, University of California at Berkeley and California Department of Transportation, Division of Structures, Sacramento, California.

Jones, R.M., and I.A. Schroeder. 1991. The Design and Construction Aspects of the Structure Research Tests at Cypress Viaduct. In Lifeline Earthquake Engineering: Proceedings of the Third U.S. Conference. Report: Technical Council on Lifeline Earthquake Engineering, Monograph 4, American Society of Civil Engineers, New York, August.

Kay, G. 1991. Structural Analysis of the Cypress Street Viaduct Test Section. Report: UCRL-ID-106870, Lawrence Livermore National Laboratory, Livermore, California, January.

Kelly, J.M., I.D. Aiken, and P.W. Clark. 1991. Response of Base-Isolated Structures in Recent Earthquakes. Structures Congress "92", Compact Papers, American Society of Civil Engineers, New York.

Lam, I.P., G.R. Martin, and R. Imbsen. 1991. Modeling Bridge Foundations for Seismic Design and Retrofitting. In Proceedings, Third Bridge Engineering Conference at Denver, Colorado, March.

Lam, I.P., and G.R. Martin. 1991. Geotechnical Considerations for Seismic Design and Retrofitting of Highway Bridges. In Proceedings, 1992 Annual TRB Meeting, Washington D.C., January.

LeBeau, R., G. Fung, A. Goldschmidt, and J. Belvedere. 1971. Post Earthquake Investigation of Bridge Damage in the San Fernando Earthquake of February 9, 1971. Report to the State Highway Engineer, Sacramento, California, March.

Lysmer, J., R.B. Seed, C.M. Mok, and S.E. Dickenson. 1991. Response of Soft Soils during the 1989 Loma Prieta Earthquake and Implications For Seismic Design Criteria. In Proceedings, Pacific Conference on Earthquake Engineering. Auckland, New Zealand. November.

Mahin, S.A., et al. 1992. Evaluation of the Seismic Performance of Retrofit Concepts for Double Deck Reinforced Concrete Viaducts. In Seminar Proceedings, Seismic Design and Retrofit of Bridges. Earthquake Engineering Research Center, University of California at Berkeley and California Department of Transportation, Division of Structures, Sacramento, California.

Mahin, S.A., and J. P. Moehle. 1990. Full-Scale Tests on the Cypress Viaduct. U.S.-Japan Workshop on Seismic Retrofit of Bridges. Proceedings: Public Works Research Institute, Tsukuba Science City, Japan, December.

Mahin, S.A. 1991. Overview of Berkeley Research and Proof Tests of Double Deck Viaduct Retrofits. Column Retrofit Tests, Proceedings of the First Annual Seismic Research Workshop, California Department of Transportation, Sacramento, California.

Maragakis, E.M., and M.S. Saiidi. 1991. Evaluation of Seismic Response of Bridges with Hinge Restrainers. In Proceedings of the First Annual Seismic Research Workshop, California Department of Transportation, Sacramento, California.

Maroney, B., and J. Gates. 1990. Seismic Risk Identification and Prioritization in the CALTRANS Seismic Retrofit Program. In Proceedings: 59th Annual Convention, Structural Engineers Association of California, Sacramento, California, September.

Maroney, B., K. Romstad, B. Kutter, M. Griggs, and E. Kasper. 1991. Experimental Measurements of Bridge Abutment Behavior. In Proceedings of the First Annual Seismic Research Workshop, California Department of Transportation, Sacramento, California.

Maroney, B., K. Romstad, and B. Kutter. 1993. Experimental Testing of Laterally Loaded Large Scale Bridge Abutments. In Proceedings: ASCE Structures Congress XI, Irvine, California, April.

Mason, J.A. 1993. Tension Pile Test. In Proceedings: ASCE Structures Congress, University of California at Irvine, April.

Mayes, R.L. 1992. Seismic Isolation Design of Bridges. In Seminar Proceedings, Seismic Design and Retrofit of Bridges, Earthquake Engineering Research Center, University of California at Berkeley and California Department of Transportation, Division of Structures, Sacramento, California.

Mellon, S., et al. 1993. Post Earthquake Investigation Team Report of Bridge Damage in the Loma Prieta Earthquake. In Proceedings: ASCE Structures Congress XI, Irvine, California, April.

Miranda, F., and V.V. Bertero. 1991. Evaluation of the Failure of the Cypress Viaduct in the Loma Prieta Earthquake. Bulletin of the Seismological Society of America, Volume 81, No. 5, Special Issue, October.

Mitchell, J.K. 1992. Ground Treatment for Seismic Stability of Bridge Foundations. Seminar Proceedings, Seismic Design and Retrofit of Bridges, Earthquake Engineering Research Center, University of California at Berkeley and California Department of Transportation, Division of Structures, Sacramento, California.

Moehle, J.P. 1992. Evaluation and Rehabilitation of Multi-Level, Multi-Column Concrete Freeways. In Seminar Proceedings, Seismic Design and Retrofit of Bridges, Earthquake Engineering Research Center, University of California at Berkeley and California Department of Transportation, Division of Structures, Sacramento, California.

Moehle, J.P., and M. Aschheim. 1992. Shear Strength and Deformability of RC Bridge Columns Subjected to Inelastic Cycle Displacements. Report No. UCB/EERC-92/04 to CALTRANS, Division of Structures, March.

Moehle, J.P., S. Mazzone, and C.R. Thewalt. 1991. Cyclic Response of RC Beam-Column Knee Joints. Report No. UCB/EERC-91/14 to CALTRANS, Division of Structures, Sacramento, California, October.

Mualchin, L., and A.L. Jones. 1992. Peak Acceleration From Maximum Credible Earthquakes in California (Rock and Stiff Soil Sites). California Division of Mines and Geology, Open File Report 92-1.

Powell, G.H. 1992a. Effective Use of Nonlinear Analysis in Bridge Design. Draft Report: to California Department of Transportation, Division of Structures, Sacramento, California, September.

Powell, G.H. 1992b. Observations on the Practical Application of Nonlinear Analysis. In Seminar Proceedings, Seismic Design and Retrofit of Bridges, Earthquake Engineering Research Center, University of California at Berkeley and California Department of Transportation, Division of Structures, Sacramento, California.

Price, T., C.B. Crouse, and R. Mitchell. 1992. Evaluation of Methods to Estimate Pile Foundation Stiffnesses for Bridges.

Priestley, M.J.N., and F. Seible. 1990. San Francisco Double Deckers—Observed Damage and a Possible Retrofit Solution. In U.S.-Japan Workshop on Seismic Retrofit of Bridges, Proceedings: Public Works Research Institute, Tsukuba Science City, Japan, December.

Priestley, M.J.N., F. Seible, and Y.H. Chai. 1991a. Flexural Retrofit of Circular Reinforced Bridge Columns by Steel Jacketing, Colret—A Computer Program for Strength and Ductility Calculation. Report No. SSRP-91/05 to the CALTRANS Division of Structures, October.

Priestley, M.J.N., F. Seible, and Y.H. Chai. 1991b. Flexural Retrofit of Circular Reinforced Bridge Columns by Steel Jacketing, Experimental Studies. Report No. SSRP-91/05 to the CAL-TRANS Division of Structures, October.

Priestley, M.J.N., F. Seible. R. Yang, J. Ricles, D. Liu, and R.A. Imbsen. 1991c. The Whittier Narrows 1987 Earthquake: Performance, Analysis, Repair, and Retrofit of the I-5/I-605 Separator. Report No. SSRP-91/08, University of California at San Diego.

Roberts, J.E. 1990. Planning For Natural Disaster Response. Presentation Outline Only: Conference on Emergency Planning, National Highway Safety Foundation, Reno, Nevada, April, 1990, City of Los Angeles; Conference on Emergency Planning and Preparedness, Los Angeles, California, March 29, 1992. Information Outline and Slides available at Division of Structures, California Department of Transportation, Sacramento, California.

Seed, R.B., S.E. Dicksenson, and C.M. Mok. 1992. Recent Lessons Regarding Seismic Response Analyses of Soft And Deep Clay Sites. In Seminar Proceedings, Seismic Design and Retrofit of Bridges, Earthquake Engineering Research Center, University of California at Berkeley and California Department of Transportation, Division of Structures, Sacramento, California.

Seible, F., and M.J.N. Priestley. 1991. Performance Assessment of Damaged Bents after the Loma Prieta Earthquake. Pacific Conference on Earthquake Engineering, New Zealand National Society for Earthquake Engineering, Wellington.

Seible, F., M.J.N. Priestley, C.T. Latham, and T. Terayama. 1993. Full Scale Test on The Flexural Integrity of Cap/Column Connections With Number 18 Column Bars. Report No. TR-93/01 to CALTRANS, January.

Seible, F., M.J.N. Priestley, N. Hamada, and Y. Xiao. 1992a. Test of a Retrofitted Rectangular Column Footing Designed to Current CALTRANS Retrofit Standards. Report No. SSRP-92/10 to CALTRANS, November.

Seible, F., M.J.N. Priestley, N. Hamada, Y. Xiao, G.A. MacRae. 1992b. Rocking and Capacity Test of Model Bridge Pier. Report No. SSRP-92/06 to CALTRANS, August.

Selna, L.G., and L.J. Malvar. 1987. Full Scale Experimental Testing of Retrofit Devices Used For Reinforced Concrete Bridges. Report No. UCLA/EQSE-87/01, Presented to California Department of Transportation, Division of Structures, June.

Sheng, L.H., and A. Gilbert. 1991. California Department of Transportation Seismic Retrofit Program the Prioritization and Screening Process. Lifeline Earthquake Engineering: Proceedings of the Third U.S. Conference, Report: Technical Council on Lifeline Earthquake Engineering, Monograph 4, American Society of Civil Engineers, New York, August.

Sitar, N., and R. Salgado. 1992. Behavior of the San Francisco Bay Mud From the Marina District in Static and Cyclic Simple Shear. In Seminar Proceedings, Seismic Design and Retrofit of Bridges, Earthquake Engineering Research Center, University of California at Berkeley and California Department of Transportation, Division of Structures, Sacramento, California.

Sweet, J. 1993. Implementation of Advanced Soil-Structure Interaction Techniques for the Analysis of Bridge Structures. Report to Division of Structures, California Department of Transportation, Sacramento, California, January.

Tamura, K., and H.C. Shah. 1991. Local Site Effects on Ground Motions at Cypress Street Viaduct Due to Loma Prieta Earthquake of 1989. In Proceedings of the Fourth International Conference on Seismic Zonation, Earthquake Engineering Research Institute, Oakland, California.

Thewalt, C.R., and B.I. Stojadinovic. 1992. Behavior and Retrofit of Outrigger Beams. In Seminar Proceedings, Seismic Design and Retrofit of Bridges, Earthquake Engineering Research Center, University of California at Berkeley and California Department of Transportation, Division of Structures, Sacramento, California.

Wilson, J.C., and B.S. Tan. 1990. Bridge Abutments: Assessing Their Influence on Earthquake Response of Meloland Road Overpass. ASCE Journal of Engineering Mechanics, Volume 116, No. 8.

Wilson, J. 1993. Modeling Bridge Abutment Stiffness on the Meloland Road Overcrossing. In Proceedings: ASCE Structures Congress XI, Irvine, California, April.

Yashinsky, M. 1990. Performance of Retrofit Measures on Existing Older Bridges. In Proceedings: 59th Annual Convention, Structural Engineers Association of California, Sacramento, California, September.

Yashinsky, M. 1992a. Post Earthquake Investigation Team Report of Cape Mendocino Earthquake. CALTRANS, Sacramento, California.

Yashinsky, M. 1992b. CALTRANS' Use of Nonlinear Analysis for the Seismic Design of Bridges. In Seminar Proceedings, Seismic Design and Retrofit of Bridges, Earthquake Engineering Research Center, University of California at Berkeley and California Department of Transportation, Division of Structures, Sacramento, California.

Zafir, Z., et al. 1990. Soil/Foundation Behavior at the Cypress Overpass (I-880) in the San Francisco Earthquake. In Proceedings of the 1990 Annual Symposium on Engineering Geology and Geotechnical Education, held at Idaho State University, Pocatello, Idaho, April.

GENERAL REFERENCES

Ang, A.H.-S., W. Kim, and S. Kim. 1993. Damage Estimation of Existing Bridge Structures. In Proceedings: ASCE Structures Congress XI, Irvine, California, April.

Astaneh-Asl, A. 1991. Behavior of Major Steel Bridges During October 17, 1989 Earthquake, Abstract Only. Structures Congress "91", American Society of Civil Engineers, New York.

Astaneh-Asl, A. 1992. Seismic Behavior and Retrofit of the San Francisco-Oakland Bay Bridge. Structures Congress"92", Compact Papers, American Society of Civil Engineers, New York.

Astaneh-Asl, A., B. Bolt, G. Fenves, J. Lysmer, and G. Powell. 1991. Seismic Condition Assessment of the Bay Bridge. In Proceedings of the First Annual Seismic Research Workshop, California Department of Transportation, Sacramento, California.

Astaneh-Asl, A. 1992. Seismic Evaluation and Retrofit of the Bay Bridge. In Seminar Proceedings, Seismic Design and Retrofit of Bridges, Earthquake Engineering Research Center, University of California at Berkeley and California Department of Transportation, Sacramento, California.

Astaneh-Asl, A. 1992. Steel Bridge Design and Evaluation Considerations Related to Local and Member Stability. In Seminar Proceedings, Seismic Design and Retrofit of Bridges, Earthquake Engineering Research Center, University of California at Berkeley and California Department of Transportation, Sacramento, California.

Astaneh-Asl, A., and J. Shen. 1993. Rocking Behavior and Retrofit of Tall Bridge Piers. In Proceedings: ASCE Structures Congress XI, Irvine, California, April.

Buckle, I.G. 1991. Revisions to the AASHTO Seismic Design Criteria for Bridges in the United States. Pacific Conference on Earthquake Engineering, New Zealand National Society for Earthquake Engineering, Wellington.

Buckle, I.G. 1991. Seismic Design Criteria for Highway Bridges. Third Bridge Engineering Conference, Report: Third Bridge Engineering Conference at Denver, Colorado, March, 1991. Report: Transportation Research Board Record 1290. TRB, National Research Council, Washington DC.

Buckle, I.G. 1990. The Preliminary Screening of Bridges for Seismic Retrofit. In Proceedings, Second Workshop on Bridge Engineering Research in Progress, University of Nevada, Reno.

CALTRANS. 1992. First Annual Seismic Research Workshop. Proceedings: California Department of Transportation, Sacramento, California.

Chai, Y.H., M.J. Priestley, F. Seible. 1991. Seismic Retrofit of Bridge Columns by Steel Jacketing. Third Bridge Engineering Conference at Denver, Colorado, March, 1991. Report: Transportation Research Board Record 1290. TRB, National Research Council, Washington, D.C.

Dickenson, S.E., and R.B. Seed. 1991. Correlations of Shear Wave Velocity and Engineering Properties for Soft Soil Deposits in the San Francisco Bay Region. Report No. UCB/EERC-91/xx, Earthquake Engineering Research Center, University of California, Berkeley, California.

Dickenson, S.E., R.B. Seed, and C.M. Mok. 1992. Recent Lessons Regarding Seismic Response Analyses of Soft and Deep Clay Sites. Seismic Design and Retrofit of Bridges, Seminar Proceedings, Earthquake Engineering Research Center, University of California at Berkeley and California Department of Transportation, Division of Structures, Sacramento, California.

Fenves, G.L., F.C. Filippou, and D.T. Sze. 1991. Evaluation of the Dumbarton Bridge in the Loma Prieta Earthquake. In Proceedings of the First Annual Seismic Research Workshop, California Department of Transportation, Sacramento, California.

Ghose, A., R.M. Polivka, and B.A. Maroney. 1991. Evaluation of Techniques for the Seismic Modeling of Elevated Freeway Bridges. Lifeline Earthquake Engineering: Proceedings of the Third U.S. Conference, Report: Technical Council on Lifeline Earthquake Engineering, Monograph 4, American Society of Civil Engineers, New York, August.

Gross, J.L., and S.K. Kunnath. 1992. Application of Inelastic Damage Analysis to Double-deck Highway Structures. Report: NISTIR-4857, National Institute of Standards and Technology, Gaithersburg, Maryland, June.

Idriss, I.M., and J.I. Sun. 1992. SHAKE 91 - A Computer Program for Conducting Equivalent Linear Response Analysis of Horizontally Layered Soil Deposits. Users Manual, University of California, Davis.

Imbsen, R.A., and R.A. Schamber. 1991. Training Program for Implementation of Newly Developed Guidelines for Seismic Design and Retrofitting of Highway Bridges. Third Bridge Engineering Conference at Denver, Colorado, March, 1991. Report: Transportation Research Board Record 1290. TRB, National Research Council, Washington DC, 1991.

Imbsen, et al. 1990. Seismic Design of Highway Bridges, Training Course Participant Workbook. FHWA Project, Second Printing, July.

Kasai, K., W.D. Liu, and V. Jeng. 1992. Effect of Relative Displacements Between Adjacent Bridge Segments. SMIP 92: Seminar on Seismological and Engineering Implications of Recent Strong-Motion Data. Proceedings: Strong Motion Instrumentation Program, California Division of Mines and Geology, Sacramento, California.

Ketchum, M.A., and C.T. Seim. 1990. Golden Gate Bridge Seismic Evaluation. Prepared in Cooperation with Imbsen and Assoc., Inc. and Gerspectra, Inc. San Francisco, California, November 2.

Ketchum, M.A., and C.T. Seim. 1991. Golden Gate Bridge Seismic Retrofit Studies. Prepared in Professional Collaboration with Imbsen and Assoc., Inc. and Geospectra, Inc. San Francisco, California, July 10.

Lysmer, J., and Deng, N. 1991. Two-Dimensional Site Response Analysis. In Proceedings of the First Annual Seismic Research Workshop, California Department of Transportation, Sacramento, California.

Maroney, B., and J. Gates. 1991. Seismic Risk Identification and Prioritization in the CALTRANS Seismic Retrofit Program. In Proceedings of the 4th U.S.-Japan Workshop on Earthquake Disaster Prevention for Lifeline Systems, Los Angeles, California.

Mayes, R.L., et al. 1990. Enhancing the Seismic Performance of Toll Road Bridges. In Proceedings: 59th Annual Convention, Structural Engineers Association of California, Sacramento, California, September.

McCallen, D., and G.L. Goudreau. 1990. Post Loma Prieta Earthquake Initiative, Seismic Analysis of an Elevated Portion of the Bay Bridge Distribution System Structure. Report: UCRL-ID-104179, Lawrence Livermore National Laboratory, Livermore, California, June 15.

Moehle, J.P., S.A. Mahin, and R. Stephen. 1990. Implications of Nondestructive and Destructive Tests on the Cypress Street Viaduct Structure. May.

Moehle, J.P. Undated. San Francisco Earthquake, October 17, 1989, Evaluation of the Performance of the I-880 Cypress Viaduct. Abstract of Research in Progress.

Mualchin, L. 1990. Seismic Hazards of CALTRANS Facilities After the Loma Prieta Earthquake and the Implications for CALTRANS Seismic Programs. International Symposium on Safety of Urban Life and Facilities: Lessons Learned from the 1989 Loma Prieta Earthquake, November 1-2, 1990. Tokyo Department of Environmental Engineering, Graduate School at Nagatsuta, Tokyo Institute of Technology, Yokohama, Japan, 1990.

Niazy, A., S. Masri, and A. Abdel-Ghaffar. 1993. Analysis of the Seismic Records of a Suspension Bridge. In Proceedings: ASCE Structures Congress XI, Irvine, California, April.

Penzien, J., W.S. Tseng, and M.S. Yang. 1991. Seismic Performance Investigation of the Hayward-BART Elevated Section Instrumented Under CSMIP. SMIP 91: Seminar on Seismological and Engineering Implications of Recent Strong-Motion Data, Preprints, Strong Motion Instrumentation Program, California Division of Mines and Geology, Sacramento, California.

Penzein, J. 1993. Seismic Design Criteria for Transportation Structures. In Proceedings: ASCE Structures Congress XI, Irvine, California, April.

Priestley, M.J.N., and F. Seible. 1992. Performance Assessment of Damaged Bridge Bents after the Loma Prieta Earthquake. Bulletin of the New Zealand National Society for Earthquake Engineering, Volume 25, No. 1, March.

Priestley, M.J.N., and F. Seible. 1993. Assessment and Testing of Column Lap Splices for the Santa Monica Viaduct Retrofit. In Proceedings: ASCE Structures Congress XI, Irvine, California, April.

Ramey, M.R., et al. 1991. Experimental Testing of Epoxy Injected and Steel Shell Retrofitted Sections from the Collapsed Struve Slough Bridge. In Proceedings of the First Annual Seismic Research Workshop, California Department of Transportation, Sacramento, California.

Rashid, Y.R., R.A. Dameron, and I.R. Kurkchubasche. 1992. Predictive Analysis of Outrigger Knee-Joint Hysteresis Tests: A Torsion/Flexure Tests at University of California, San Diego. Report to University of California, Berkeley, University of California, San Diego, and CAL-TRANS, Sacramento, California.

Roberts, J.E. 1989. Bridge Seismic Retrofit Program For California Highway System. In Proceedings: U.S.-Japan Workshop on Lifeline Earthquake Engineering, Public Works Research Institute, Tsukuba Science City, Japan, May.

Roberts, J.E. 1989. Theory of California Seismic Bridge Design And Analysis For The Beginner. California Department of Transportation, Division of Structures, Training Course for Entry Level Engineers, July.

Roberts, J.E. 1990. Recent Advances in Seismic Design and Retrofit of Highway Bridges. In Proceedings: Earthquake Engineering Research Institute Annual Meeting Palm Springs, California, May.

Roberts, J.E. 1990. Recent Advances in Seismic Design and Retrofit of Highway Bridges. Structural Engineers Association of California, Proceedings: 59th Annual Meeting, Incline Village, Nevada, September.

Roberts, J.E. 1991. Recent Advances in Seismic Design and Retrofit of Highway Bridges. Proceedings: Seismic Retrofit Workshop, University of California at San Diego, July.

Roberts, J.E. 1991. Seismic Performance of Steel Bridges. Structural Steel Fabricators Annual Meeting, Saint Louis, Missouri, September, 1991. Published in Modern Steel Construction magazine, July 1992.

Roberts, J.E. 1991. Recent Advances in Seismic Design and Retrofit of Highway Bridges. In Proceedings of the Third U.S. Conference, Report: Technical Council on Lifeline Earthquake Engineering, Monograph 4, American Society of Civil Engineers, New York, August.

Roberts, J.E. 1991. Seismic Retrofitting of San Francisco Viaducts. In Proceedings: 60th Annual Meeting, Structural Engineers Association of California, Palm Springs, California, October.

Roberts, J.E. 1992. Seismic Design of Bridge Foundations. In Proceedings, Transportation Research Board Annual Meeting, Washington, D.C., January.

Roberts, J.E. 1992. Large Scale Model Testing for Seismic Design and Retrofit, Initiating, Funding and Managing. Presented at the Earthquake Engineering Research Institute Annual Meeting, San Francisco, California, February.

Roberts, J.E. 1992. Research Based Bridge Seismic Design and Retrofit Program, Criteria, Standards and Status. In Proceedings: Fifth U.S.-Japan Workshop on Earthquake Disaster Prevention for Lifeline Systems, Tsukuba Science City, Japan, October 26.

Roberts, J.E. 1992. Effect of Foundation Soil Response on Bridge Seismic Performance. Proceedings: Fifth U.S.-Japan Workshop on Earthquake Disaster Prevention for Lifeline Systems, Tsukuba Science City, Japan, October 26.

Roberts, J.E. 1992. Bridge Seismic And Other Research Needs-CALTRANS Overview. In Proceedings: Third NSF Workshop on Bridge Engineering Research in Progress, University of California at San Diego, November 15.

Roblee, C.J. 1992. Synthesis of CALTRANS' Foundation Seismic Research Program. Internal Report: Division of New Technology, Materials & Research, California Department of Transportation, Sacramento, California, November.

Saadeghvazri, M.A. 1990. Response of the Struve Slough Bridge under the Loma Prieta Earthquake. In Proceedings: Second Workshop on Bridge Engineering Research in Progress, University of Nevada at Reno.

Schnabel, P.B., J. Lysmer, and H.B. Seed. 1972. SHAKE, A Computer Program for Earthquake Response Analysis of Horizontally Layered Sites. Report: Earthquake Engineering Research Center, University of California, Berkeley.

Seim, C., and S. Rodriguez. 1993. Seismic Performance and Retrofit of the Golden Gate Bridge. In Proceedings: ASCE Structures Congress XI, Irvine, California, April.

Sikorsky, C., N. Stubbs, and M. Richardson. Nondestructive Damage Assessment of a Bridge Using Modal Testing and Structural Reliability.

Stuart, R.J. 1991. Seismic Modeling Parametric Studies. In Proceedings of the First Annual Seismic Research Workshop, California Department of Transportation, Sacramento, California.

Thorkildsen, E. 1992. Overview of CALTRANS' Bridge Seismic Research Program. Structure Notes, No. 27, California Department of Transportation, Sacramento, California, July.

Werner, S.D., and C.E. Taylor. 1990. Seismic Risk Considerations for Transportation Systems. Recent Lifeline Seismic Risk Studies, Report: Technical Council on Lifeline Earthquake Engineering, Monograph 1, American Society of Civil Engineers, New York, November.

Werner, S.D., S.A. Mahin, and N.C. Tsai. 1993. Compilation and Evaluation of Current Bridge Damping Data Base. Report to CALTRANS, February.

Werner, S.D., L. Katafygiotis, and J. Beck. 1993. Seismic Analysis of Meloland Road Overcrossing Using Calibrated Structural and Foundation Models. In Proceedings: ASCE Structures Congress XI, Irvine, California, April.

Whittaker, A.S., E. Elsesser. 1991. Seismic Design Criteria for Transportation Structures. Lifeline Earthquake Engineering: Proceedings of the Third U.S. Conference, Report:Technical Council on Lifeline Earthquake Engineering, Monograph 4, American Society of Civil Engineers, New York, August.

Zelinski, R. 1990. California Highway Bridge Retrofit Strategy and Details. In Proceedings: 59th Annual Convention, Structural Engineers Association of California, Sacramento, California, September.

Zelinski, R. 1990. California Highway Bridge Retrofit Strategy and Details. In Proceedings: Second Workshop on Bridge Engineering Research in Progress, University of Nevada at Reno.

Zelinski, R. 1991. San Francisco Double Deck Viaduct Retrofits. Lifeline Earthquake Engineering: Proceedings of the Third U.S. Conference, Report: Technical Council on Lifeline Earthquake Engineering, Monograph 4, American Society of Civil Engineers, New York, August.

Zelinski, R., and A.K. Dubovik. 1991. Seismic Retrofit of Highway Bridge Structures. Lifeline Earthquake Engineering: Proceedings of the Third U.S. Conference, Report: Technical Council on Lifeline Earthquake Engineering, Monograph 4, American Society of Civil Engineers, New York, August.

COMMISSIONED RESEARCH PROJECTS

R-1. Retrofitting of Bridge Columns, Stage I, using third to half scale model tests. UC, San Diego. Contract $400,000. Completed June 30, 1990. This is the first contract with Dr. Priestley which was intitated in early 1987. The emphasis was on single columns using steel jackets for confinement. The physical testing is complete and all reports are completed.

R-2. Guidelines for Effective use of Nonlinear Structural Analysis for Bridge Structures. UC, Berkeley. Contract $ 47,000. Completion July, 1993. Principal investigator Dr. Graham Powell. This project is for providing guidelines and training to the Department in the use of the new non-linear analysis programs recently installed. Work is underway.

R-3. Shear Strength Capacity Vs. Rotation of Column Pins at Base of Elevated Roadway Structure, using third to half scale model tests. UC, Irvine. Contract $80,000. Completion date April 30, 1993. Principal Investigator Dr. Robin Shepard. This project is necessary because of the many columns which are pinned at the base on multi-column bents. As we move into the multi-column phase of seismic retrofit, we need this information. Report will be submitted spring 1993.

R-4.   Retrofitting of Bridge Columns, Stage II, using third to half scale model tests. UC, San Diego. Contract $331,000. Completion June 30, 1991. This is the second contract with Dr. Priestley and is a continuation of the previous work using different retrofit techniques and different column types. Work is completed and reports are submitted.

R-5.   Experimental Testing of Epoxy Injected Steel Shell Retrofitted Sections From Collapsed Struve Slough Bridge, using full-scale tests. UC, Davis. Contract $53,000. Completion date June 30, 1993. Principal investigator Dr. Melvin Ramey. We salvaged several broken piles from the Struve Slough Bridge at Watsonville for this test. Since we have a large number of these type bridges in the state, both state and locally owned, it is necessary to test techniques for improving their seismic performance. Work completed. Reports will be submitted spring 1993.

R-6.   Evaluation of the Dumbarton Bridge in the Loma Prieta Earthquake. UC, Berkeley. Contract $50,000. Completion date June 30, 1991. Principal investigator Dr. Gregory Fenves. The Dumbarton Bridge was instrumented with strong motion instruments and the records are available. It is the only long crossing of the San Francisco Bay which was designed to modern bridge earthquake codes and criteria and this performance evaluation will be useful to the profession. Work completed. Report submitted.

R-7.   Seismic Response of Deep Soil Sites in the San Francisco Bay Area. UC, Berkeley. Contract $315,000. Completion date December 31, 1992. Principal investigators Dr. John Lysmer, Dr. Raymond Seed. As a major part of the comprehensive earthquake vulnerability evaluations of important transportation structures we have a need to first determine the foundation response. This research is a direct result of problems with structures constructed on deep, soft soils during the Loma Prieta earthquake. The results of this project will be a new set of Acceleration Response Spectra (ARS curves) for deep, soft soils. Work completed. Report due spring 1993.

R-8.   Retrofitting of Bridge Columns, Stage III, using third to half scale model tests. UC, San Diego. Contract $768,000. Completion July 12, 1993. This is the third contract with Dr. Priestley and is a continuation of the previous work using different retrofit techniques and different column types. This project, however, concentrated on the multiple column bridge bent configuration where the columns typically have moment connections at both top and bottom, but in many cases are pinned at the base. Work completed. Report due spring 1993.

R-9.   Seismic Retrofit of Bridge Column Footings, using third to half scale model tests. UC, San Diego. Contract $374,000. Completion July 12, 1993. This is the fourth contract with Dr. Priestley and is the final contract in the series to evaluate and test the best techniques to retrofit the footings of older bridges where the original design moments introduced into the footings were much smaller than now anticipated after columns are retrofitted to carry more moment. Work underway.

R-10.   Evaluation and Retrofitting of Multi-Level and Multi-Column Structures, using third to half-scale model tests. UC, Berkeley. Contract $1,900,000. Completion December 31, 1993. Principal investigator, Dr. Stephen Mahin. This research is essential for the future evaluation of older and newer multi-level structures. Interim results will be used to confirm the techniques and details being used in the current retrofit program. Work underway.

R-11.   Develop Bridge Seismic Design Criteria with Higher Degree of Safety and Reliability than Provided with Current Design Procedures. Applied Technology Council. Contract $800,000. Completion date is December 15, 1993. This project is to evaluate the current CALTRANS/ AASHTO bridge design criteria/code and recommend improvements and changes.

R-12.   Evaluation of the Performance of Bridge Cable Restrainers During the Loma Prieta Earthquake. University of Nevada, Reno. Contract $91,000. Completion date June, 1993. Principal investigator, Dr. M. Saiidi. Work complete. Final report due spring 1993.

R-13. Reduced Scale Tests of Pier Walls Under Cyclic Loading for Seismic Retrofit. UC, Irvine. Contract $461,000. Completion date July 1992. Principal investigator Dr. Robin Shepard. Because of the many bridge piers of this design we need test results to determine the best retrofit technique and to provide better design criteria for new pier wall designs. Work completed. Report due spring 1993.

R-14. Development of High Strength Fiber Composite Column Wrap. Fyfe Associates, Inc. Contract $73,000. Completion date December, 1991. This is an alternative method to provide confinement on existing bridge columns. It has been used in Japan for industrial smoke stacks using carbon fiber, which is ten times the cost of other fiber composites. Work complete. Report submitted.

R-15. Inspection and Data Collection on the Substructure of the Bay Bridge. UC, Berkeley. Contract $49,900. Completion date not yet negotiated. The principal investigator, Dr. Astaneh needed additional funding to complete work begun under an NSF grant. Work complete. Report included with major contract on SFOBB.

R-16. Flexural Integrity of Column/Cap Connections Using Number 18 bars. UC, San Diego. Contract $500,000. Completion date December 1992. The principal investigator, Dr. Priestley will build and test full size components to test the bond development length required for these large bars. Most large bridge columns in California utilize these bars and there is some question regarding the adequacy of the AASHTO code requirements. Work complete. Draft report submitted.

R-17. Seismic Evaluation of the San Francisco-Oakland Bay Bridge. GENESYS. Proposed contract amount $160,000. This proposal is for a rapid evaluation of the west bay spans using current technology. Has been evaluated by the Seismic Research Advisory Panel. Contract negotiations underway.

R-18. Evaluation of Earthquake-Induced Cyclic and Permanent Ground Displacements for Soil Sites. Earth Mechanics. Proposed contract amount $50,000. Still being evaluated.

R-19. Development and Implementation of Improved Seismic Design and Retrofit Procedures for Bridge Abutments. University of Southern California. Proposed contract amount $100,000. Completion date not yet determined. Still being negotiated.

R-20. Experimental Measurements of Bridge Abutment Stiffness and Strength Characteristics. UC, Davis. Contract amount $350,000. Completion date June, 1993. Principal investigator, Dr. Karl Romstad. This project will incorporate the centrifuge to test models of various combinations of bridge abutment-soil interaction. Work still underway. Report due late spring 1993.

R-21. A Simplified, Verified Procedure to Analyze Soil-Pile Structure Interaction During Earthquake Loading Conditions Using an Effective Stress Method. UC, Davis. Contract amount $25,000. Completion date December 1991. Principal investigator Dr. I.M. Idriss. Draft Report due spring 1993.

R-22. Response of Pile-Supported Bridge Elements due to Liquefaction. National Cooperative Highway Research Program (NCHRP). Contract cost will be borne by NCHRP if this project is approved. It has national application. Being evaluated by research committee.

R-23. Seismic Condition Assessment of the Bay Bridge. UC, Berkeley. Contract amount $800,000. Contract completion date June 30, 1993. Principal investigator, Dr. Abduhollah Astaneh. This project is the major effort to model and conduct an extensive dynamic analysis of the bay bridge with time history to evaluate its response to a larger earthquake such as an 8.0 on the San Andreas or a 7.3 on the Hayward fault. We have decided to begin with the San Francisco-Oakland Bay Bridge for obvious reasons. This project will include a comprehensive analysis of the foundation material response and upon completion we plan to retrofit the bridge to withstand the forces and movements recommended. The project is the beginning of the assessment of all major bay and river crossings in California. Work completed. Draft report submitted. Final report due June 1993.

R-24. Seismic Hazard Risk Analysis for the San Francisco Bay Region. This proposed project will evaluate the seismic hazard risks for the region and provide appropriate ground acceleration input for geotechnical evaluations of specific structure sites. Geomatrix Consultants completed work. Final report delivered February 1993.

R-25. Seismic hazard risk analysis for Southern California. This proposed project will evaluate the seismic hazard risks for the region and provide appropriate ground acceleration input for geotechnical evaluations of specific structure sites. Consultant, Woodward Clyde Associates. Contract completion date, June 1993. Work completed. Draft report due June 1993.

R-26. Seismic Condition Assessments of Remaining Toll Bridges. These projects will be advertised by the RFP process and appropriate principal investigators and consultants will be selected for the 9 remaining major toll crossings on the state highway system. They are the Dumbarton Bridge, the San Mateo-Hayward Bridge, The Richmond-San Rafael Bridge, the Carquinez Bridge (1927 and 1955 structures), the Benicia-Martinez Bridge, the Antioch Bridge, the Terminal Island Suspension Bridge (Vincent Thomas Bridge), the Gerald Desmond Bridge on Terminal Island (soon to be taken into the state highway system) and the San Diego-Coronado Bridge.

R-27. Implement Advanced Soil Structure Interation Techniques for the Analysis of Bridge Structures. Consultant, Coast Analytics, Incorporated. Contract completion April 1993. Contract amount $88,500. Work underway. Report due May 1993.

R-28. Response of Soft Soil Sites Using the Centrifuge. UC Davis. Various Site Conditions will be Tested at Varying Levels of Shaking to Augment Recordings From Recent Earthquakes and to Calibrate Analytic Procedures. Contract Amount $125,000. Contract Completion Date October, 1993. Principal Investigator Dr. I.M. Idriss.

R-29. Construction of Shaker forthe Large Centrifuge. U.C. Davis. Partial Support for Construction of This Shaker for Future Research Involving Soil-Pile and Soil-Structure Interaction, Liquifaction, Site Improvement and other projects. Contract Amount $82,500. Contract Completion Date November, 1993. Principal Investigator Dr. I.M. Idriss.

DISCUSSANTS' COMMENTS: BRIDGES

James D. Cooper, Federal Highway Administration

We owe a lot to CALTRANS for the way they reacted to the earthquake—namely, in their openness in allowing investigators to come in, examine, and quantify the damage and in their dissemination of information to researchers. This is the chief reason we learn from earthquakes.

My interest stems from what the lessons from the Loma Prieta earthquake mean nationally. For example, in California, there are 24,000 bridges; to date, 1,300–1,400 of those have been retrofitted. Nationally, there is an inventory of 577,000 bridges with as many as 75 percent of those being bridges at risk, either because of location or no design for seismic resistance, and few have been retrofitted.

We have learned lessons from previous earthquakes—the San Fernando event being one of the most significant events for the bridge community. Lessons from previous events point to the need to define and accommodate forces, accommodate displacements, evaluate ground-motion amplification and attenuation, and identify liquefaction potential from site-specific studies or from macro analysis. We have also learned that retrofit enhances performance.

Loma Prieta was a moderate earthquake, with a short duration of strong ground motion. It produced two to three cycles of inelastic response. We need to consider carefully whether the earthquake can be used as a base for design of structures in other areas of the country, particularly those east of the Rockies, or whether eastern events will produce larger cycles of inelastic response. The latter will have significant impact on the design/retrofit philosophy.

Nevertheless, there were technical lessons learned from the Loma Prieta event:

• *Simple retrofit helps.* Following the San Fernando earthquake, CALTRANS embarked on a program to identify vulnerable bridges and details and implement a simple, relatively inexpensive retrofit technique to provide displacement control across expansion joints. The relatively good performance of those bridges retrofit with hinge restrainers is testimony that large numbers of structures can be economically retrofit to enhance seismic resistance.

• *Detailing is critical.* Bridges designed and constructed following the San Fernando earthquake performed relatively well. Column and column-connection details were revised to accommodate earthquake-generated shears, moments, and pull-out forces. In addition, the development of improved analytical technology was key to improved detail design.

• *Vulnerability assessment is required.* Identification of hazard exposure, coupled with site and structural analyses are required to determine structural vulnerability. Only when the system as a whole is compared can a rational decision be made as to which structures—and to what extent—to retrofit.

Mitigation Strategy

• *Design new structures to current criteria.* The increased cost associated with providing seismic resistance in newly designed and constructed bridges can vary significantly. However, typical cost increases are about 2–3 percent, a very affordable figure. If all new (or replacement) bridges were designed by the latest standards, it would take upwards of 100 years to reduce the seismic vulnerability of the highway system. In the long term, this strategy is affordable and will prevail.

• *Retrofit the most important/critical bridges.* Since most retrofit is very costly, retrofit only those structures that are defined as important with simple, relatively inexpensive technology. Retrofit only those bridges that are defined as critical with the more complex (thereby costly) technology as it evolves.

• *Multidisciplinary approach required.* The Loma Prieta earthquake taught us that scientific and engineering knowledge is advancing; many public policy, legal, and financial issues remain to be resolved; public consciousness and awareness about the catastrophic consequences of a great earthquake are being raised, and emergency planning and response procedures are advancing. It is clear that a sound earthquake-hazard-mitigation strategy will require the coordinated and cooperative involvement of professionals of varied disciplines.

Initiatives Required

• *Awareness.* Continue promoting technical and public awareness programs.
• *Evaluation.* Complete seismic evaluation of the bridge inventory.
• *Design update.* Develop philosophically consistent design criteria.
• *Retrofit criteria.* Develop and adopt a rational retrofit criteria.

In conclusion, earthquake-hazard-mitigation is a long-term endeavor. As we implement new technologies, we are making a significant impact on improving the seismic performance of our highway system.

Gregory Orsolini, Parsons DeLeuw, Inc.

The I-280 Southern Freeway viaduct in San Francisco is a project I am intimately familiar with, having been working on it since just days after the Loma Prieta earthquake. This is a structure on a very high acceleration site and the poorest soils in the Bay Area. We were asked to adhere to the basic philosophy of CALTRANS—to prevent collapse—but also to have a serviceability mechanism addressed. The project was to replace the columns, the joint areas, and the foundations. I'll be concentrating on the reconstruction of these joints and outriggers (some in excess of 50 ft in length).

Beam-Column Joint Shear Design

Current design practice for beam-column joint shear for bridges results in heavily reinforced connections. Regarding beam-column joints in bridges, there are some unique considerations—the tendency to use some sizeable bar sizes (#14 and #18 are not uncommon) in retrofit and new construction, and to use large joint shear stirrups. This leads to complications for placement because of the consideration of the bend diameters. We are still understanding/learning about density of reinforcement in the joints.

In building these joints, we are fighting uncertainty in the construction industry about how these things are put together. As the contractors learn what we're trying to do, the learning curve should accelerate quickly. To get adequate implementation, we are interacting with contractors and inspectors on how this is actually done.

Bent Cap Outrigger Design

The original design used a column that was pinned at the top and the bottom, which reduced the lateral bending in these outrigger caps to zero. Our peer review panel asked us to increase redundancy. We did that by adding a fixed joint at the bottom, which adds lateral bending problems to the design of outriggers. That is where we got the flared configuration.

For vertical loads, we have post-tensioning in a parabolic shape but also post-tensioning following a bit of a flare. We added bolsters within the box girder to take the reaction of those large flares.

We approached the combination of vertical and lateral bending by trying to keep the strains and the pre-stressing strains to below the proportional limit (.008 strain). To consider the vertical accelerations, we multiplied the vertical dead-load by 50 percent—either increasing or decreasing the vertical load by that amount.

Use of Simplified Nonlinear Analysis Methods for Seismic Analysis

There are many options for nonlinear analysis available. I would like to encourage the use of a more simplified type of analysis, as it has a lot of benefits and is a good tool to use.

A number of useful nonlinear analysis methods are currently being used for assessing the ductile behavior of existing structures—displacement ductility, equal energy, and equal displacement concepts can be applied to bridge-retrofit work. It does not necessarily have to be a complex procedure to use nonlinear analysis.

Thank you very much.

Nicholas F. Forell, Forell/Elsesser Engineers

After the Loma Prieta earthquake, it was determined that all of the six San Francisco double-deck freeways were in need of retrofit. Most of the freeways (the exceptions were the Alemani Viaduct and the Terminal Separation structure) were damaged and closed to traffic. These structures were built in the late 1950s and have similar construction to the Cypress Viaduct, which failed catastrophically.

Immediately after the earthquake on November 1, 1989, CALTRANS retained six major engineering firms with adequate staffing to immediately design and implement repairs and retrofits. The Governor's Board of Inquiry, in its hearings, concluded that because there was no precedent for this type of retrofit, CALTRANS should initiate a peer review process to assure compliance of the designs with the performance criteria established.

The peer review panel selected by CALTRANS consisted of six practicing engineers experienced in seismic design and four technical advisors.

The technical advisors (two professors from UC Berkeley and two from UC San Diego) were extremely important because of their in-depth knowledge of analysis and concrete design as well as their research experience. The technical advisors supplemented the technical knowledge of the panel members and the consultants.

An early and detailed establishment of the scope of work of the peer review panel is important. In this case, the panel reviewed the seismic design criteria and the geologic-hazard report, the design and performance criteria, applicable and available research results, analysis and modeling assumptions, design details, construction documents, and constructability. The panel did not perform a check on the drawings' calculations; therefore peer review cannot be considered a substitute for independent plan checks.

It is most important that the peer review panel be convened at the very beginning of the project.

Although the consultants were on board immediately after the earthquake, the peer review panel, for many reasons, could not be convened until March 1990. This resulted in wasted effort and money. Some of the work under construction had to be halted or abandoned and much of the design redone. It is therefore important to have the peer review panel functioning as the retrofit concepts are formed.

On unique and complex projects, the peer review process can be laborious. On the double-deck retrofit projects, 56 hearings were held. Yet, the peer review process proved itself invaluable and provided CALTRANS with the assurance that the retrofit design will meet the established performance criteria.

John Clark, Anderson Bjornstad Kane Jacobs

I would like to address what *worked*. We've seen a great deal of spectacular things that failed, but it is important to address what did work—anything designed to the current code (i.e., the post-1983 guidelines that Mr. Cooper alluded to—the ATC-6 document).

The ATC-6 document was a revelation that we are dealing with a displacement phenomena. Under the old—even the post-1971 San Fernando codes—we simply doubled our forces and went about things the same way. The code we have now is a good code. I don't mean to say that it solves all our problems. The current code has brought a strong need to consider the foundation effects—the liquefaction potential. Unfortunately, in building bridges, we don't have the luxury to choose the best sites. Consideration should be given to soil-structure interaction effect—not only as it may tend to amplify things but also as it may tend to reduce the response through damping and other effects. The strength of the current code is its concentration on providing ductile details and in providing adequate seat length. Mr. Cooper made an important point about the efficacy of simple retrofits—the joint restrainers. The empirical design tools we have for that are very crude, but they work.

Seismic isolation is another very promising tool. There was not anything in the bridge field during the Loma Prieta earthquake that gave us a good test, but it is a good principle. There are things we need to learn about it yet—particularly the increased vulnerability at the joints and the means to address that.

We also need to focus our research on whether we can relax some of the confinement provisions we have in the code. That's a constructability issue more than anything else. There are tools out there that we need to get into our code—such as a reduction factor based on our percentage of axial load. We have not yet incorporated that into the code because people think they are being conservative. I would beg to differ with that.

A few brief comments on what I might call the retrofit "philosophy": There is a critical need to think clearly about what we are doing in codifying the retrofit process in general, because it is a very cost-benefit-sensitive issue. The need for retrofitting far outweighs the available resources. Do decision makers (legislators and the general public) know the risk involved, and are they willing to accept it? For consultants, is this codified to protect us from our legal brethren? As Greg Orsolini pointed out, we need to codify or provide more guidance on how to use the simple nonlinear techniques and to focus on system ductility as opposed to member component ductility and strength. Finally, I would ask researchers to help in the field. How do we really design for the long-duration effects? Are the attenuation relationships the same in these long-duration earthquakes? Are there frequency shifts due to more rapid attenuation of the higher frequencies? These are some future points that we critically need some help with.

Thank you.

7

Recovery, Mitigation, and Planning

George G. Mader

It is indeed a humbling experience to attempt to write a keynote address on the broad subject I have been assigned. All of us attending have impressions of what was learned from the Loma Prieta earthquake, but how much do we know in detail about all of the lessons that can be gleaned from the event. What worked and what didn't work? Seven other speakers will address this same topic as discussants of this paper or in a panel this afternoon. Some of these individuals were directly involved in the recovery process after the earthquake, and others have carried out intensive research projects focused on aspects of the recovery. They can speak with more authority than I with regard to many aspects of the topic. I am certain they will add much to my comments and will probably provide different points of view and also corrections to what may be erroneous statements on my part. With these qualifications, I will now embark into dangerous waters.

I will attempt to do two things. First, I will try to provide at least an overview of what I believe to be seven important topical areas in which a considerable amount was learned from this earthquake. Second, I will try to emphasize those lessons that may not be covered by others addressing this same subject area. The topics I will address, listed in order are funding, recovery plans and regulations, mitigation actions, state laws, demolition, housing, and business recovery.

Of the various reports and papers I have read regarding the Loma Prieta earthquake that address the practical aspects of rebuilding after the event, I have to single out the California Seismic Safety Commission's report *Loma Prieta's Call to Action*, as an outstanding contribution (California SSC, 1991). The commission held eight hearings on the earthquake from October 1989 to January 6, 1990. These hearings were held in Sacramento and in the locales where major damage occurred including San Francisco, Oakland, Santa Cruz, Watsonville, and San Jose. In addition, the communities of San Francisco, Oakland, Santa Cruz, Watsonville and Los Gatos and the county of Santa Cruz were invited to prepare their own recommendations for improving policies and programs. These recommendations are included at the end of the commission's report. The comments appear to represent views valid as of the early part of 1991, something over a year after the event. I will rely extensively on this report for my paper.

FUNDING

My impression of the major recurring theme in the community reports that are contained in the Seismic Safety Commission's document is of the extreme frustration experienced by public and private interests in obtaining financial assistance critical to their recovery and reconstruction. This is certainly not news, but the vital role of assistance in recovery cannot be denied. Also, the intense problems of funding agencies in administering programs cannot be underestimated. Those administering programs are put in the unenviable position of having to see that funds are spent according to dictates of programs. Not only do they have to see that funds match detailed requirements, they also have to make certain that claims do not falsely describe the conditions of damage. They are open to intense criticism because they often have to say "no."

I do not pretend to be an authority on funding programs, but the comments by jurisdictions need to be considered. San Francisco in its report stated:

> The time involved for individuals applying for individual assistance was sometimes excessive because of the large number of different agencies and regulations involved. For example, because of confusion about the scope of the authority and responsibility between the Office of Emergency Services (OES), Federal Emergency Management Agency (FEMA), and the Department of Housing and Urban Development (HUD) and the Department of Transportation, it took nine months to determine whether expenditures to repair a possible landslide affecting private homes, public land and streets were eligible for reimbursement, and by what agency. . . . They are still displaced at the time of this writing—a year and a half after the earthquake.

San Francisco's report details many other examples of concern. For instance, it states that since the California Natural Disaster Assistance Program (CALDAP) funds are available only to those who have been denied federal aid, it was necessary to go through the process of being denied by a federal agency, even though the outcome was a forgone conclusion, in order to be considered for CALDAP aid. The report indicates this delay of up to six months simply discouraged many from applying. The report also states that frustration with federal programs was so high that the U.S. representative for the district had to assist over 200 constituents through the process.

Other jurisdictions had other concerns. For instance, the city of Santa Cruz cited the fact that FEMA aid is for replacing "in kind" what was lost. The city points out that where improvements were inadequate to begin with, there needs to be funding for "betterment" through some kind of a loan program, perhaps by the state. Many other criticisms have been made. For example, it is reported that estimators for FEMA, often from another part of the country, were unfamiliar with local building costs and therefore did not properly value the cost of rebuilding.

The horror stories regarding obtaining financial assistance are legendary. I believe that no one thinks it is a simple problem and that most would agree that funding agencies try to carry out a difficult task in a fair and reasonable manner. Given this very brief glimpse of some of the types of problems that occurred, what practical lessons have been learned? Or phrased another way, what suggestions have been put forward for improvements? Let us consider a few.

1. The report of Los Gatos indicates that those who had earthquake insurance were the ones who, in general, were most successful in obtaining financing to rebuild. If this is representative, it appears that additional attention to earthquake insurance is still an avenue that needs to be pursued by individuals and at the national level. Something akin to flood insurance appears needed.

2. Oakland indicates that in most instances FEMA field representatives were from out of town. The city recommends that FEMA adopt a policy of assigning the same personnel to work with the jurisdiction on a regular basis so that familiarity is established. Also emphasized was the need to have representatives fluent in the languages of residents of the city.

3. San Francisco pointed out that it had no prior experience or preparation in dealing with federal disaster response. The city then ". . . hired one full-time, experienced federal recovery expert to train and coordinate other City staff who were to be involved in federal and State programs. About twenty City staff people devoted full time to these programs for about nine months." This highlights the need for local jurisdictions to have at least one staff member who is up-to-date on federal and state funding programs so that the jurisdiction will have its own expert when disaster strikes. Such a person would be invaluable when the jurisdiction is confronted with complex funding programs at the same time the city is in chaos after an earthquake.

4. The problems with funding programs provide an argument for considering major changes in how they operate. San Franciso has suggested, for instance, that the "Administration of the federal Public Assistance Program could be considerably simplified by changing the character of the program from one of reimbursement for exhaustively detailed expenses, to one that distributes 'block' grants to local agencies. State and federal agencies could establish granting

criteria which measure the magnitude of a disaster . . . and the estimated damage. Assistance would occur quickly and without a detailed application process. Local governments could establish their own priorities for short-term and long-term recovery assistance and then make a full accounting of expenditures" (California SSC, 1991).

At the Bay Area Earthquake Preparedness Project (BAREPP) conference "Putting the Pieces Together" (BAREPP, 1990), held one year after the earthquake, Dianne Guzman, Planning Director for Santa Cruz County during the earthquake, made what she called two "critical recommendations" with respect to the repair and replacement of housing. One of her recommendations states: "There is a need for clearer expectations on the level of long term state and federal assistance for all aspects of repair and rebuilding of housing." This comment could be extended to buildings other than housing.

Based on this brief review of the funding problem, the following lessons have been summarized:

1. Consideration needs to be given to major simplifications in the federal and state funding programs.

2. Funding programs need to be more clearly defined so as to provide guidance to local communities for long-term recovery.

3. FEMA needs to be better oriented to the communities it is going to assist.

4. Local jurisdictions need to have in-house experts in federal and state funding programs.

5. Insurance has proven effective in getting money to the insured and needs to be pursued as a major recovery tool.

I should add that even with what I have written, I have only addressed the tip of the iceberg. The matter of funding is fundamental to the recovery and rebuilding process and requires much additional consideration. The Seismic Safety Commission's report states: "In short, the disaster assistance application and review processes needs a thorough overhaul."

RECOVERY PLANS AND REGULATIONS

A manager of the firm that contracted with Santa Cruz County to provide repair and reconstruction permitting services summed up the challenge to government when he said:

> Governments should focus on and formalize the process they want to have in place when disaster strikes rather than trying to reinvent or tweak or get around or ignore normal procedures when you're trying to move as quickly as possible to help your community.

California SSC, 1991

In the Earthquake Engineering Research Institute (EERI) journal devoted to the Loma Prieta earthquake, the problem was further emphasized with the statement:

> The lack of recovery planning in all jurisdictions is glaringly obvious. There are no preplanned programs, and all decisions appear ad hoc and characterized by linear thinking rather then systematic approaches.
>
> Benuska, ed., 1990

It is clear that after an earthquake the normal ways of doing business are not adequate to accommodate both the pressure for speed in approving projects and the volume of applications. Also, the event brings to light the defects in the environment that have been ignored over the years, such as areas of ground failure and weak buildings. As a result, the jurisdiction is faced with not only the damage but, at that late date, attempting to impose stricter regulations than had been in place prior to the earthquake. To property owners, this amounts to putting a greater burden on the unfortunate. Most of the experience seems to point to a need to consider the recovery and rebuilding process long before the earthquake. Indeed, the report of the California Seismic Safety Commission recommends that the state should require local jurisdictions to develop community recovery plans with at least as much emphasis as is placed on emergency response plans. Given this perspective, what are some specific recommendations based on the lessons from the Loma Prieta earthquake?

Some special organizational structure and guidelines appear necessary. In Los Gatos, one of the first actions of the city council after the earthquake was to establish the Earthquake Restoration Committee and policies governing demolition, repair, and reconstruction of damaged buildings and unreinforced masonry. The council also established procedures for allowing use of a motor home or trailer as temporary housing.

Los Gatos and Santa Cruz both adopted similar policies with respect to reconstruction of buildings. In Los Gatos, if a building was to be replaced with another building identical in size and use as previously existed and for which no more than 100 square feet of floor area were to be added, the approval process was simplified and normal fees were not required. On the other hand, if the replacement structure was to be substantially different or more than 100 square feet larger, it had to go through the regular development review process, and fees were not waived. This policy backed up the council's desire to get the town back to the condition it was in prior to the earthquake.

Problems inherent in changing regulations, especially when the changes are more restrictive, are pointed out by Santa Cruz County's report. It states that existing policies allowed response with relatively few surprises but that when new technical information indicated a need to modify policies and make them more restrictive, major difficulties arose. Most notably this occurred in the Santa Cruz Mountains where large landslides occurred. Proposals by county staff to prevent residents from repairing and rebuilding homes in this area became politically explosive. Dealing with the uncertainties of large landslide areas in the calm of a non-disaster is hard enough, but it is easier than dealing with them in the aftermath of a disaster. Policies established in advance of the earthquake to govern reconstruction in such areas would have been very useful in the aftermath of the earthquake. Such high risk areas and related policies need to be in place well before the damage occurs. Given the intense pressures local jurisdictions are under after an earthquake to allow rebuilding in hazardous areas, the Santa Cruz County report goes further and recommends state support in establishing regulations to govern building in hazardous areas and also in financial measures to help solve difficult situations through engineering remedies and relocation of development.

Organizational approaches to directing the rebuilding of damaged areas varied in cities. Los Gatos appointed the Earthquake Restoration Committee already mentioned. The city council of Santa Cruz, about two months after the earthquake, appointed a 36-member citizen group, Vision Santa Cruz, to develop an overall concept for rebuilding the downtown. The city council of Watsonville appointed a Downtown Recovery Committee. While the nature of problems differed between the three communities, two moved ahead rather rapidly with putting things back as they were, while Santa Cruz undertook an extended debate with community involvement to determine what type of a rebuilt downtown was desired. Some thought in advance of a major earthquake as to the organizational structure to put in place to guide the rebuilding would seem to pay off in the aftermath.

One of the rather surprising aspects of the earthquake was not the result of direct damage but rather the major community controversies that arose indirectly from the damage in several areas. The two most notable examples are the Cypress Structure failure and the Embarcadero Freeway failure. In the case of the Cypress Structure, the community through which the freeway passes opposed its reconstruction, as they argued it divided the community and was not compatible with the residential uses in the area. In the case of the Embarcadero Freeway, those who had long considered the structure to be a blight on the San Francisco waterfront argued that the structure should not be replaced. In each instance, there followed periods of intense controversy and major changes to plans for rebuilding. The new solutions are generally perceived to be more consistent with local desires.

The foregoing observations and many others that relate to this topic indicate a need to carefully consider the rebuilding process that a community will have to go through after a major earthquake, but such consideration needs to be well in advance of the event. Based on the comments of the cities, the following lessons appear to be evident.

1. Well in advance of an earthquake, local jurisdictions need to establish overall programs for handling the recovery and rebuilding that will take place

after the earthquake. These programs need to have a number of components as suggested below.

2. The organizational structure needed to guide intensive rebuilding efforts needs to be defined and the rules for its operation established. Options seem to include the elected body itself, a committee, or possibly a redevelopment agency. Also, staff needs for the body need to be considered and met.

3. Ordinances need to be prepared that will become effective automatically upon declaration by the governing body of the jurisdiction. These ordinances should cover a range of topics, some of which are

    a. applications deemed routine should be exempted from time-consuming reviews and normal fees;

    b. permission should be given to use recreation vehicles, campers, etc., for normal living for short periods;

    c. provisions should be established for temporary use of non-commercial buildings by displaced businesses; and

    d. provisions should be made on how to handle repair and rebuilding in areas that experience ground failure.

4. Any plans for organizational structure and procedures to be followed after an earthquake need to take into consideration that unforeseen controversy might arise. This argues for carefully anticipating where major damage might occur and then evaluating the various potential impacts on the rebuilding process.

5. As an assist in developing the organization and procedures to respond to an earthquake, earthquake scenarios might be used by those who would be involved in rebuilding to help them understand the various problems with which they will be confronted. In this way, they would be better able to be prepared. This approach is similar to that used by emergency-response personnel but aimed at the first several years after an earthquake rather than the first several days.

6. Arrangements should be made to obtain additional personnel and expertise to handle the large increase in applications for repairing and replacing structures. This pertains to building inspectors and engineering geologists especially.

MITIGATION

The major lesson with respect to mitigation is that where mitigation steps were taken, damage seems to have been light. One can't help being impressed by how much went right in the earthquake rather than how much went wrong. Most buildings stood up, and relatively few land failures caused damage to structures. The failures that occurred are well documented and by and large were the result of old buildings built to inadequate standards or buildings located in potentially hazardous areas, such as areas subject to landslides, liquefaction, and enhanced ground shaking. Most of these problems could have been solved through better land-planning decisions or engineering. The major structural failures, such as the freeways already mentioned and the Bay Bridge, were mostly due to engineering that was not equal to the task. In short, we know how to evaluate land for construction better than we did in the past; we know how to engineer solutions to ground failure better than we did; and we know how to design buildings better than we did. The problem, of course, is that much of the environment was built before the state of the art and practice had advanced to where they are now. If this is the case, then what are the concerns at this juncture in history?

The major challenge is to use current knowledge in an effective manner. In those jurisdictions where there is a lag, practice needs to be brought up to date. The major lessons include the following:

1. Hazardous buildings should be strengthened or removed, especially unreinforced masonry buildings.

2. Old houses should be bolted to foundations, shear bracing added to cripple walls where needed, and mobile homes properly tied down.

3. Lands should be planned and zoned so as to avoid hazardous areas, or solutions to provide safety should be engineered and provided.

4. Subdivisions and building permits should be subjected to rigorous requirements for geologic information, and that information should be reviewed by qualified geologists on behalf of local jurisdictions.

None of the foregoing suggestions are new or startling; however, they are not uniformly carried out. These are minimal requirements if significant improvements are to take place. Carrying these out probably requires state legislation, a topic addressed next in this paper.

IMPORTANCE OF STATE LEGISLATION

Patricia Bolton and Carlyn Orians, in their report *Earthquake Mitigation in the Bay Area, Lessons from the Loma Prieta Earthquake* , address the question of whether the earthquake caused local jurisdictions to take a more aggressive approach in mitigating earthquake hazards (Bolton, 1992). Their conclusions, based on evaluations of 39 jurisdictions in seven counties, was that by and large no significant advancements took place. They point out that some jurisdictions reported a heightened awareness of earthquake hazards and that over one-third "indicated increased attention to emergency response preparedness, but not to loss reduction measures." Since the aftermath of an earthquake is generally deemed to be the most likely time to obtain agreement for improvements in earthquake safety, this is perhaps a discouraging commentary.

On the other hand, what was the record at the state level after the Loma Prieta earthquake? The California Seismic Safety Commission reports that 443 bills in the 1989–1990 legislative session and the 1989 extraordinary session included the terms "seismic safety" or "earthquake" and that 137 were signed into law by the governor. In the 1991–1992 legislative session, an additional

139 bills were introduced, and 55 were signed by the governor into law. All but eight of these bills can be broken into the following groups (California SSC, 1992):

Category 1, Existing Vulnerable Facilities	61
Category 2, New Facilities	24
Category 3, Emergency Management	22
Category 4, Disaster Recovery	61
Category 5, Research and Information/Education	16

When one considers the major advancements in seismic safety throughout California, it becomes clear that most of the actions that have been taken in the hazard-mitigation area are a direct result of state legislation. Bolton and Orians point out that the most common strategies both before and after the earthquake with respect to mitigation were application of the Uniform Building Code for Seismic Zone 4; the inclusion of a safety element in the general plan, which identifies seismic hazards and recommends programs for reducing hazards; and conformance with the California Unreinforced Masonry (URM) Law (SB 547).

There are instances, on the other hand, where local jurisdictions have made improvements in hazard mitigation through changes in planning requirements or through an increased awareness of the reality of earthquake hazards, which causes them to pay more attention to the problem. Bay Area Earthquake Preparedness Project (BAREPP) and the Southern California Earthquake Preparedness Project (SCEPP) have been major catalysts in enhancing the level of awareness and providing mitigation instruction and informative publications. These efforts take considerable time to pay off, but on balance have been, in this writer's opinion, almost "the only game in town," in these areas. On balance, however, it appears abundantly clear that the general level of mitigation throughout the state is largely a result of state-mandated requirements. Political pressures at the local level against imposing increased requirements in controlling the use of land and buildings are intense. The state legislature, however, is more isolated from these parochial pressures. As Bolton and Orians point out, the shifting of the political pressure from the local level makes a difference. They state: "Once a measure is required by state law, adoption of the program becomes more of an implementation issue rather that a political issue." So, what are the practical lessons learned from this experience?

1. Continued focus must be placed on developing appropriate state legislation that will further advance earthquake-hazard mitigation at the local level. Local governments cannot be left to their own devices.

2. Even with state-mandated requirements, local jurisdictions should be encouraged to develop local programs attuned to their particular needs. The past work of BAREPP and the SCEPP made major strides in this area, and these types of efforts should be continued.

DEMOLITION

The handling of demolition was a major problem facing cities that experienced significant amounts of building collapse or partial collapse in the earthquake. Considerable experience was gained, and some important insights were developed with respect to various aspects of the problem. That demolition is a significant concern is emphasized by the fact that of the seven major recommendations from local governments with respect to community recovery (California SSC, 1991), two address the problems of demolition, and they recommend the following:

1. Standards are needed for the repair of damaged buildings, including historical buildings.

2. Criteria are needed for the demolition of severely damaged buildings, including historic buildings.

On the surface, the question of whether a building should be demolished might seem rather simple. Is it more economical for the owner to demolish the building or rebuild it? The problem immediately becomes more complex as the owner begins to face the question of what the jurisdiction may demand, or allow, and what financing might be available. The entire demolition process, from the first building evaluations to the more detailed ones and finally to decisions to demolish or repair is indeed complex. One aspect, however, stood out in the earthquake—the question of historic buildings.

The EERI Loma Prieta Reconnaissance Team (Benuska, 1990) points out that while the percentage of buildings considered historic that were damaged in the earthquake was small, these buildings were individually significant and were often grouped in older areas where they tended to constitute a historic area. To illustrate the small numbers, consider that in the of city of Santa Cruz, 53 of the 1,025 damaged buildings, or 5 percent, were classified as historic. Probably, the major controversy in the city in the aftermath of the earthquake was over the fate of the historic St. George Hotel.

The hotel, while listed in the city's Historic Building Survey and named as a contributing building in a historic district, was not listed on the National Register of Historic Places. Several groups, Friends of St. George, the California Preservation Foundation, and the National Trust for Historic Preservation, filed a suit to prevent demolition after the city council had authorized the demolition being sought by the building owner. The basis of the suit was lack of compliance with a bill passed just after the earthquake (SB3x) and the California Environmental Quality Act. The city's decision was based on an engineer's report that the building presented an imminent threat to public safety. Ultimately, the court upheld the city's decision, but in the interim, the building remained and delayed planning and rebuilding for an important part of the badly damage downtown area (Cronin, undated).

As pointed out in the EERI report (Benuska, 1991), one FEMA rule tends to act against preservation. The rule provides that FEMA will pay for demolition that takes place within 30 days of the earthquake. After that time, the building owner must pay for the demolition. It is argued that this time pressure does not allow adequate time to properly evaluate the building and the various options for preservation. As pointed out in the report: "Owners need to be made aware of the importance of their historic buildings and encouraged to repair rather than demolish them." One of the ancillary aspects of this topic is the need for building standards to which historic buildings must be restored, a topic presumably addressed by others at this symposium.

What are some of the practical lessons learned in this subject area?

1. Cities and counties, before an earthquake, should have in place a well-defined program for assessing damage and making determinations as to demolition decisions, especially with respect to historic buildings.

2. Standards for repair of damaged buildings, especially historic ones, should be in place well before an earthquake.

3. Building owners and local jurisdictions should be well aware of the impact historic building designations will have on reconstruction after earthquakes or other disasters and consider the appropriateness of retaining the buildings.

HOUSING

It is estimated that approximately 24,000 residential structures sustained damage in the Loma Prieta earthquake. In Santa Cruz County, approximately 14,100 housing units were damaged, and of these about 900 were destroyed. Thus, of the county's 90,000 housing units, 16 percent were affected. In Oakland, approximately 1,000 units were destroyed, while in San Francisco about 500 were permanently lost and an additional 500–600 were severely damaged and were not restored as of early 1991 (California SSC, 1991).

It appears that most of the damaged housing units have been repaired using federal loans, conventional loans, or insurance payments. The majority of the affected owners and residents have been rehoused even though it may have taken a considerable amount of time. The major problem, and one that remains, is handling the needs of the low-income population, the population that typically occupies old housing of poor construction, most notably URM hotels and apartment buildings. The U.S. General Accounting Office, assigned to investigate the adequacy of the federal response to the earthquake, noted that the Loma Prieta earthquake was the first large-scale disaster in a major urban area where the problem of repairing or replacing low-income housing occurred (California SSC, 1991). The problem was compounded by the fact that in the Bay Area there was a large homeless population and an inadequate supply of housing for low-income persons before the earthquake.

Repair and reconstruction of low-income rentals is not normally feasible using conventional financing. The cost of repairs typically results in increased rents, which exceed the resources of the low-income occupants. After the earthquake, FEMA took the position that it could not provide funds to restore or replace damaged units. Neither did HUD take on the responsibility. The SSC report states very clearly: "The basic conclusion related to damaged-housing replacement is that federal disaster-assistance programs do not provide adequate assistance to state and local governments to reconstruct damaged rental units." Only Small Business Administration (SBA) loans were available, but they were not adequate. In order to help address this problem, the state had previously established the California Natural Disaster Assistance Program for Rental Properties after the Whittier Narrows earthquake to provide low-interest, deferred loans for rehabilitating rental units. The record indicates, however, since these loans are a measure of last resort, assistance was very slow and favored nonprofit housing groups (California SSC, 1991).

What is the practical lesson learned from this experience? The SSC report puts the problem as well as it can be stated:

> The State of California urgently needs to focus on the issue of low cost housing replacement especially when contemplating response to the expected urban earthquakes that are likely to permanently displace tens of thousands.

Also, as has already been mentioned:

> There is a real need to solve the easier problems including tying down mobile homes, bolting old houses to their foundations, and adding shear strength to cripple walls of old houses. Also, there is the need to solve the more difficult problems of URM residential buildings.

BUSINESS RECOVERY

A first order question is: How did the Bay Area business community fare as a result of the earthquake? The Association of Bay Area Governments report *Macroeconomic Effects of the Loma Prieta Earthquake* provides some answers to this question (ABAG, 1991). The report leads off with the statement that the "The Loma Prieta earthquake produced minimal disruption to the overall economy of the Bay Area and its environs." This assertion is backed up with research on employment and taxable retail sales. A few key statistics will suffice to make the point. The estimated loss of 7,100 jobs, when extended for a four-month period amounts to a loss of less than a quarter of one percent of the economy, which has more than three million jobs. The report also points out, however, that the effects are felt differently in the region. For instance, San Francisco lost approximately $73 million in taxable sales in the fourth quarter of 1989. The report attributes this loss to damage to transportation and infrastructure. In Santa Cruz County, unemployment claims jumped from 3,910 to 7,246 for the period between the third week of October and the second week of November in 1989 as compared with 1988. It is clear, therefore, that while the region faired rather well, certain areas felt the economic impact to a much larger extent.

Richard Wilson, City Manager of Santa Cruz, in his report *The Loma Prieta Quake, What One City Learned* (Wilson, 1991), described the vulnerability of local economies very clearly:

> A disaster does not fundamentally change the larger economy; it simply takes out the most vulnerable parts of the pre-disaster economy. The economic trends that were in place before the disaster will continue. The major change will be that the economic activities lost in the disaster will be resumed elsewhere.

He describes the damage as follows:

> For all practical purposes, Santa Cruz's downtown—the source of about 20 percent of the city's sales taxes—was destroyed by the earthquake. The city's two department stores, at either end of the downtown mall, had to be demolished, as did thirty other buildings. Few of the buildings that remained were fit for occupancy, and in any event access could not be gained to the streets and sidewalks adjacent to buildings.

> Some of the businesses, those that could afford to do so, moved to locations outside of Santa Cruz, where space was available. Some businesses, the more marginal ones, simply ceased to exist. Others moved into some empty buildings near the downtown, and still others were relocated into rapidly erected "pavilions" on downtown parking lots. After the initial clean-up and the commencement of the use of temporary quarters, the replanning of the downtown started with the appointment of the 36-member Vision Santa Cruz. Today, the downtown is slowly being rebuilt.

Based on this intense and instructive experience, Mr. Wilson lists a number of practical lessons that have general applicability, including the following:

1. The first priority is to protect against the immediate loss of economic activity to other places. This requires that pre-disaster plans identify locations for temporary business structures such as public parking lots and private properties.

2. Pre-disaster plans should identify key parties that will represent business, government, and community interests in the aftermath of an earthquake, and such groups should meet periodically.

3. Pre-disaster recovery planning should contemplate the potential financial realities that will be faced and should be prepared in advance. This can include having local expertise in federal and state aid programs as well as other sources of financing.

4. A governmental structure should be defined that will take charge of managing the rebuilding process. While the exact expected damage cannot be forecast and therefore a detailed plan for rebuilding cannot be prepared, "the framework within which a city will work can be defined."

CONCLUSIONS

In the area of mitigation, the Loma Prieta earthquake appears to have reconfirmed that basic urban-planning techniques, when they take into consideration seismic factors, can provide a reasonable level of safety from ground failure. This is based on the observation that most ground failures occurred in areas developed in the past when the recognition of seismic problems in the urban-planning process was less. Large areas of ground failure in new developments were not recorded. On the other hand, some of the newer developments may not have been adequately tested simply because the intensity of ground shaking at the site and the duration of shaking were not sufficient.

Also, in the area of mitigation, the earthquake reconfirmed what has been known for a long time, that older, weak buildings fail unless they have been properly reinforced. The earthquake did point out, however, that an adequate job of carrying out such reinforcement activities is not being done. Even this was known before the earthquake, as local policy makers refused to face up to the problem.

As suggested in this paper, probably the major practical lesson was what was learned about the failings of the funding process. Financial resources are the fuels that drive the rebuilding engines after an earthquake. It appears that the entire funding process needs a review, and major changes are warranted.

Also, it is clear that a number of communities learned through experience that they had to develop organizational structures, procedures, and regulations to deal with rebuilding in the heat of the post-earthquake rebuilding process. There appears to be unanimity that these matters need to be attended to prior to a major earthquake.

Unfortunately, it appears that local governments left to their own devices by and large will not adopt the seismic safety regulations necessary to protect local populations and improvements. The record seems to show that the most successful mitigation efforts were a direct result of state legislation. This again brings attention to the state level for developing and adopting the most critical items of legislation.

In the area of demolition, the problems presented by historic buildings came into sharp focus. These buildings can be both an asset and a curse. They provide historic context for a city but also can stand in the way of a coordinated rebuilding effort after an earthquake. Much more attention needs to be paid to historic designations so that on balance they play a proper role in the city and the reconstruction process.

Finally, the loss of housing emphasized the plight of low-income persons who typically occupy old, weak buildings. This provides urgency to efforts to strengthen old residential buildings and to develop better financing mechanisms to be available immediately after an earthquake.

The examples of lessons included in this paper are admittedly a selected list based on my readings and personal observations. Others should add to the list based on their perspectives. It is clear, however, that there is still a long way to go in responding to the challenges in the areas of recovery, mitigation, and planning.

REFERENCES

ABAG (Association of Bay Area Governments). 1991. Macroeconomic Effects of the Loma Prieta Earthquake. ABAG.

BAREPP (Bay Area Earthquake Preparedness Project). 1990. Proceedings: Putting the Pieces Together. BAREPP.

Benuska, L., Technical Editor. 1990. Earthquake Spectra. Supplement to Volume 6, Loma Prieta Earthquake Reconnaissance Report, May 1990.

Bolton, P.A., and C.E. Orians. 1992. Earthquake Mitigation in the Bay Area, Lesson from the Loma Prieta Earthquake. Summary Report, Batelle.

Cronin, T.G. Undated. Overview of Legal Issues Attending the Loma Prieta Earthquake. California Seismic Safety Commission.

California SSC (Seismic Safety Commission). 1992. California at Risk: Reducing Earthquake Hazards, 1992-1996. Report SSC 91-08, Seismic Safety Commission.

California SSC. 1991. Loma Prieta's Call to Action. Seismic Safety Commission.

Wilson, Richard C. 1991. The Loma Prieta Quake: What One City Learned. International City Management Association, Washington, D.C.

DISCUSSANTS' COMMENTS: RECOVERY, MITIGATION, AND PLANNING

Kenneth C. Topping, Independent Consultant

I would like to underscore everything that George Mader has just said. First, for context, I would like to expand the term "recovery" to include two parts: short-term and long-term. The words that are missing are "reconstruction" and "land-use." Ultimately, there are three major principles of reconstruction—land-use, land-use, and land-use—to paraphrase the old real estate terminology. The idea that a city can be better *after* a disaster is something that has been practiced in only a few cities around the world—such as Nagoya and Hiroshima, Japan. Yet there are examples that wise rebuilding can lead to a better community.

Here in San Francisco we have the traditional phases of the disaster. From the Presidio collection of photographs of the 1906 earthquake, there are examples of the destruction, the housing issue (emblematic of recovery stage), and the reconstruction—ceremonializing what you have at the end (the Palace of Fine Arts—the one piece from the exhibition that memorialized the earthquake and then ultimately announced that the reconstruction was over). Daniel Burnham's plan to rebuild San Francisco quite differently was rejected in the process—the monument remains, but many of the problems were recreated.

I would like to briefly focus on the contrast between that vision—of being able to make a better place and to show through physical symbols that it is better—and the reality. The reality is born out by the fact that although hazard mitigation should shape development, hazards exist in the built environment that are very substantial—not just earthquake hazards but others as well. For example, Oakland has learned from its experience of the earthquake and the fires in just four years. The built pattern dictates what can and cannot be done to a considerable degree in the mitigation area. This is an area where there is a linkage between the planners and the engineers—for issues such as how to design-in safety (as embodied in the Stafford Act, Sections 404 and 409) and how to reconcile the fact that people need to get back into their homes quickly.

From the Loma Prieta earthquake (in the Santa Cruz mountains, in particular) and from the Oakland-Berkeley Hills fire, a series of special recovery considerations have emerged. One consideration is that of the nature of the recovery organization, if any. You can plan, set in place, and adopt an ordinance that would automatically authorize by the governing body a recovery organization that would run parallel with the emergency organization. It would extend well beyond the emergency period, through the years that it takes to reconstruct a community. This is fundamental to the major change that is needed in this area of recovery and reconstruction planning. We just are not looking far enough into the future. We have seen some form of ad hoc reconstruction organization emerge in every major catastrophic situation. Mr. Mader is correct—if we learn nothing else, it is that a reconstruction organization must be planned for so that it can be activated by the declaration of the emergency.

Another critical consideration is building moratorium ordinances, along with hazard-evaluation procedures. Hazard (not damage) evaluation is critical as it has to do with whether or not permits are issued in areas where they shouldn't be. Other considerations include: temporary site uses, repair standards, and permit expediting; nonconforming buildings and uses; illegal buildings and uses; the potential for resubdivision (currently there is no way to reassemble land quickly, with private incentive, as other countries do).

Mr. Mader's point that urban planning can help in these disaster reconstruction situations fundamentally gets down to land-use planning—not just building design. It involves such things as paved road widths and fire-safety measures. In Oakland, they formed a Benefit Assessment District, which is something that ought to be looked at in relation to earthquake hazards. There is a law on the books for geologic-hazards assessment districts, but it is very cumbersome. In Oakland, the Benefit Assessment will bring in $20 million a year over the next 10 years (the sunset period), which will be used for vegetation management and fire suppression. This Assessment District just happens to straddle the Hayward fault—so there is a fundamental seismic safety issue in the replanning of this area. The replanning has to be incremental, practical, and within the scope of what is feasible in terms of overall policy.

I'd like to close with the idea that a globally applicable technology of reconstruction planning is needed. Thank you.

Patricia A. Bolton, Battelle Seattle Research Center

Yesterday we heard from Gary Johnson that there is going to be greater federal-level emphasis on a national mitigation strategy. We also heard Tom Tobin's eloquent call to advocacy—what the people in this room can do to promote earthquake mitigation. This call to advocacy recognizes that we have increasingly precise information on how to prevent damage. But it has continually proven to be difficult for local jurisdictions to get earthquake mitigation to the top of the political agenda so that agencies can commit resources to doing mitigation.

George Mader mentioned findings from my research on mitigation planning following the Loma Prieta earthquake. I would like to make three further observations based on that research and other findings and observation. Looking at the registration list, there are at least 50 officials from local-level agencies registered here. Most of the rest of you are consultants and academics. I am talking to all of you about how to use your expertise to be advocates.

I am going to talk about getting mitigation programs "off the shelf," using best sellers in mitigation planning, and "boxing up" mitigation planning. Although I am going to spare you the jargon of my discipline and address them in this somewhat light-hearted way, each of these topics is a serious matter for enabling local jurisdictions to move forward with mitigation. These topics are related to using incentives, using technical assistance, and developing organizational strategies for implementing mitigation.

The Hazard Mitigation Grant Program, included in the Staff Act, was implemented after the Loma Prieta earthquake to provide matching funds to communities for mitigation projects. Some experienced problems coming up with mitigation projects for their proposals, but our observations indicated that those who had some type of project on the shelf had an easier time with their proposal. As a community, you can ask if you have some project that was "left on the stump," that is, provided by some consultant but never acted on. You might find you already have several projects on the shelf that never got funded. Also, the Mitigation Grant Program legislation requires that you have a mitigation plan and update it annually. This is an opportunity to "rotate your stock" of mitigation projects out to the front to see what might sell at this point.

With respect to "best sellers" in mitigation-planning assistance, our research proposed a list of sources of mitigation-planning-assistance documents that had been made available over the years to California planners. Two sources came to the fore as familiar and useful to local agency staff: The Bay Area Regional Earthquake Preparedness Project assistance and technical guides and the opportunities to learn about problems and solutions through one's own professional association's seminars and material. Local officials like technical assistance that is easily accessible, continuously available, locally relevant, and provided by credible sources.

The last topic is how to package mitigation programs, or the "boxes" we found mitigation programs in when we talked to these communities. Different boxes have different consequences for what gets done. In some places, mitigation was still in the jury box—"we're still weighing the evidence for the hazard." Another situation might be referred to as the Pandora's Box—the program was viewed as "a prolific source of troubles." Then there is the sandbox—a playing around with projects with no internal structure for when or in what order we do things. Another model to consider is the Whitman Sampler Box—where you create a new box with the choice pieces out of all our other boxes. There seems to be a tremendous value in taking an interagency perspective on mitigation planning. It will take all of us to reduce earthquake losses.

Jerold H. Barnes, Salt Lake County Planning

I have learned a great deal from the Loma Prieta earthquake. When I first started in the hazards-mitigation business, I thought that this goal of ours was a big elephant that we had to eat, but we could only do that a bite at a time. This was the way were progressing. The Loma Prieta earthquake convinced us that the beast is much larger, but that our ability to take larger bites hasn't grown.

We've always looked to California for the state of the art in planning and things that we should do. In Utah, we haven't experienced a really large damaging earthquake, although the scientists tell us that we have a good chance of a 7.5 in the not so distant future somewhere along one of the segments of the Wasatch fault.

Utah is a different world regarding hazards mitigation than California. The Utah Seismic Safety Council was active from 1977–1981. When work was completed, the State Legislature was given a report with a set of recommendations for legislative needs to address. The following are two examples. (1) The state should adopt the Uniform Building Code for the whole state. That recommendation was not followed until the late 1980s—8 or 9 years later. The jury is still out on how the small communities are dealing with that building code. (2) Legislation should be adopted that would give definite authority to implement a hazard-reduction program. That recommendation took until 1990 to happen. The legislation, however, simply allows that environmental plans *may* have an environmental element, which, in turn, *may* have some geologic-hazard mapping.

Nevertheless, important work is being done through universities, the State Division of Comprehensive Emergency Management, the Utah Geologic Survey, some engineering firms, and local governments. In Utah, it is the local governments that are leading the way—not the state.

In Utah, we need to learn from your experience. If I take any of the scientific lessons that I have heard here to my county commission and say this is what we need to do, it will just go on a pile somewhere. We need to have translated information in simple terms from the scientists so that we can put that to work and bring it to the attention of those who make decisions on programs and budgets. Thank you.

APPENDIXES

A

Symposium Presenters

PLENARY SESSION KEYNOTERS

VITELMO V. BERTERO, University of California, Berkeley
G. WAYNE CLOUGH, Virginia Polytechnic Institute and State
University, Blacksburg
RONALD EGUCHI, EQE International, Irvine, California
PAUL F. FRATESSA, Paul F. Fratessa Associates, Oakland, California
GEORGE MADER, William Spangle and Associates, Portola Valley,
California
JAMES E. ROBERTS, California Department of Transportation,
Sacramento
KATHLEEN TIERNEY, University of Delaware, Newark
THOMAS TOBIN, California Seismic Safety Commission, Sacramento

PLENARY SESSION DISCUSSANTS

RICHARD ANDREWS, California Governor's Office of Emergency
Services, Ontario
DONALD BALLANTYNE, Kennedy/Jenks/Chilton, Federal Way,
Washington
JEROLD BARNES, Salt Lake County Planning Division, Salt Lake City,
Utah
JAMES E. BEAVERS, Martin Marietta Energy Systems, Oak Ridge,
Tennessee

PATRICIA A. BOLTON, Battelle Human Affairs Research Center, Seattle, Washington

JOHN H. CLARK, Anderson Bjornstad Kane Jacobs, Seattle, Washington

JAMES D. COOPER, Federal Highway Administration, McLean, Virginia

WILLIAM R. COTTON, William Cotton & Associates, Los Gatos, California

NICHOLAS F. FORELL, Forell/Elsesser Engineers, San Francisco, California

THOMAS HANKS, U.S. Geological Survey, Menlo Park, California

WILLIAM T. HOLMES, Rutherford & Chekene, San Francisco, California

GERALD JONES, City of Kansas City, Missouri

STEPHEN A. MAHIN, University of California, Berkeley

THOMAS D. O'ROURKE, Cornell University, Ithaca, New York

GREGORY ORSOLINI, Parsons-DeLeuw, Inc., San Francisco, California

MAURICE S. POWER, Geomatrix Consultants, San Francisco, California

HENRY R. RENTERIA, Office of Emergency Services, Oakland, California

CHARLES ROBERTS, Port of Oakland, Oakland, California

RICHARD D. ROSS, Missouri Emergency Management Agency, Jefferson City

THOMAS STATTON, Woodward-Clyde Federal Services, Las Vegas, Nevada

LACY SUITER, Tennessee Emergency Management Agency, Nashville

STEVEN PHILIPS, Pacific Gas and Electric Company, San Francisco, California

KENNETH C. TOPPING, Urban Planning Consultant, Pasadena, California

CONCURRENT BREAKOUT SESSION SPEAKERS

CHRISTOPHER ARNOLD, Building Systems Development, San Mateo, California

ROBERT BOLIN, New Mexico State University, Las Cruces

ROGER D. BORCHERDT, U.S. Geological Survey, Menlo Park, California

ROBERT BRIDWELL, CALTRANS District 4, Oakland, California

MARY C. COMERIO, University of California, Berkeley

CHARLES EADIE, City of Watsonville, California

WILLIAM ELLSWORTH, U.S. Geological Survey, Menlo Park, California

JAMES FINLEY, Besselberg, Keesee and Associates, Inc., San Francisco, California
SIGMUND A. FREEMAN, Wiss, Janney, Elstner Associates, Emeryville, California
NICHOLAS P. JONES, Johns Hopkins University, Baltimore, Maryland
JACK KARTEZ, Texas A&M University, College Station
LAWRENCE KORNFIELD, City of San Francisco, California
SHIRLEY MATTINGLY, City of Los Angeles, California
JACK P. MOEHLE, University of California, Berkeley
CHRIS D. POLAND, H.J. Degenkolb Associates, San Francisco, California
CHRISTOPHER ROJAHN, Applied Technology Council, Redwood City, California
SAIID M. SAIIDI, University of Nevada, Reno
WILLIAM U. SAVAGE, Pacific Gas and Electric Company, San Francisco, California
CHARLES SCAWTHORN, EQE Engineering, San Francisco, California
NICK SITAR, University of California, Berkeley
PAUL G. SOMERVILLE, Woodward-Clyde Consultants, Pasadena, California
DIANA TODD, National Institute of Standards and Technology, Gaithersburg, Maryland
FRED M. TURNER, Seismic Safety Commission, Sacramento, California
STUART D. WERNER, Dames & Moore, Oakland, California
RICHARD WILSON, City of Santa Cruz, California
T. LESLIE YOUD, Brigham Young University, Provo, Utah

SYMPOSIUM COORDINATORS

CAROLINE CLARKE GUARNIZO, Director, Board on Natural Disasters
PETER H. SMEALLIE, Director, Geotechnical Board
SUSAN TUBBESING, Executive Director, Earthquake Engineering Research Institute

APPENDIX
B

Symposium Agenda

MARCH 22, 1993 • DAY 1

7:00 am   Registration

8:00 am   Welcome
— Lloyd Cluff, *Chair, NRC Organizing Committee and President, EERI*
Robert Wesson, *USGS*
William Anderson, *NSF*
Gary Johnson, *FEMA*
Richard Wright, *NIST*

Symposium Keynote Address:

"LEGACY OF THE LOMA PRIETA EARTHQUAKE: CHALLENGES TO OTHER COMMUNITIES"
— L. Thomas Tobin, *California Seismic Safety Commission*

9:00 am   **PLENARY SESSIONS**

GEOTECHNICAL ISSUES
— G. Wayne Clough, VPI & State University, *Keynote Speaker*
— Lloyd Cluff, *Moderator*

Discussants:
William R. Cotton, *William Cotton & Associates*
Maurice Power, *Geomatrix Consultants*
Thomas Hanks, *USGS*
C. Thomas Statton, *Woodward-Clyde Consultants*

10:30 am   Break

11:00 am   BUILDINGS
— Paul Fratessa, Paul F. Fratessa Associates, Inc., *Keynote Speaker*
— Lloyd Cluff, *Moderator*
Discussants:
James Beavers, *Martin Marietta Energy Systems*
Stephen A. Mahin, *University of California, Berkeley*
Gerald Jones, *Kansas City Building Department*
William Holmes, *Rutherford & Chekene*

12:30 pm   Lunch

245

2:00 pm   EMERGENCY
           PREPAREDNESS AND
           RESPONSE
           — Kathleen Tierney, University
             of Delaware, *Keynote
             Speaker*
           — Shirley Mattingly, City of
             Los Angeles, *Moderator*
           Discussants:
           Henry Renteria, *City of Oakland*
           Richard D. Ross, *Missouri
             Emergency Management
             Agency*
           Richard Andrews, *California
             Governor's Office of
             Emergency Services*
           Lacy Suiter, *Tennessee
             Emergency Management
             Agency*

3:30 pm   Break

4:00 pm   **CONCURRENT BREAKOUT
           SESSIONS**

           GEOTECHNICAL ISSUES
           — Clarence Allen, California
             Institute of Technology,
             Moderator
           A. Geologic/Geophysical
              Studies
           — William Ellsworth, *USGS*
           B. General Description of
              Ground Motion on Rock
           — Paul Somerville, *Woodward-
             Clyde Consultants*

           BUILDINGS
           — William Holmes, Rutherford
             & Chekene, *Moderator*
           A. Performance of Building
              Codes (Function vs. Safety)
           — Chris D. Poland, *H. J.
             Degenkolb Associates*
           B. Retrofits: How Did They
              Perform?
           — Diana Todd, *NIST*

C. Timely Demolition/
   Engineering Input
— Laurence Kornfield, *City of
   San Francisco*

EMERGENCY
PREPAREDNESS AND
RESPONSE
— Thomas Beckham, South
   Carolina Office of
   Emergency Services,
   Moderator
A. Predisaster Planning
— Shirley Mattingly, *City of
   Los Angeles*
B. Search and Rescue
— Eric Noji, *National Center
   for Environmental Health,
   CDC*
C. Emergency Shelter
— Robert Bolin, *New Mexico
   State University*
D. Damage Assessment
— Christopher Rojahn, *Applied
   Technology Council*

5:30 pm   Adjourn

MARCH 23, 1993 • DAY 2

8:15 am   Introductions to Day 2
           — Lloyd Cluff

8:30 am   **PLENARY SESSIONS**

           LIFELINES
           — Ronald T. Eguchi, EQE
             Engineering & Design,
             Keynote Speaker
           — Wilfred D. Iwan, California
             Institute of Technology,
             Moderator
           Discussants:
           Thomas D. O'Rourke, *Cornell
             University*
           Donald Ballantyne, *Kennedy/
             Jenks/Chilton*

Charles Roberts, *Oakland Port Authority*
William Savage, *PG&E*

10:00 am Break

10:30 am BRIDGES
— James Roberts, CALTRANS, *Keynote Speaker*
— Ian Buckle, NCEER, *Moderator*
Discussants:
James D. Cooper, *Federal Highway Administration (invited)*
Frieder Seible, *University of California, San Diego*
Nicholas F. Forell, *Forell/ Elsesser Engineers*

11:30 am RECOVERY, MITIGATION AND PLANNING
— George Mader, William Spangle & Associates, *Keynote Speaker*
— Dennis Wenger, Texas A & M University, *Moderator*
Discussants:
Kenneth C. Topping, *Independent Consultant*
Patricia Bolton, *Battelle Research Center*
Gerald Barnes, *Salt Lake County Planning Department*

12:30 pm Lunch

2:00 pm **CONCURRENT BREAKOUT SESSIONS**

GEOTECHNICAL ISSUES
— T. Leslie Youd, Brigham Young University, *Moderator*
A. Liquefaction/Damage
— T. Leslie Youd

B. Landslides
— Nicholas Sitar, *University of California, Berkeley*
C. Soft Ground/Site Effects
— Roger Borcherdt, *USGS*

BUILDINGS
— Chris D. Poland, H. J. Degenkolb Associates, *Moderator*
A. Non-structural/Architectural Issues
— Christopher Arnold, *Building Systems Development*
B. Managing Risk and The Right to Know
— Fred M. Turner, *California Seismic Safety Commission*
C. New Standards for the Performance of Elevators
— James Finley, *Hesselberg Keesee & Assoc.*
D. Slightly Damaged Buildings
— Sigmund A. Freeman, *Wiss, Janney, Elstner Associates*

LIFELINES
— Wilfred D. Iwan, California Institute of Technology, *Moderator*
A. Performance of Underground Utilities (Gas Lines, Water, Sewer)
— Charles Scawthorn, *EQE Engineering*
B. Performance of Transportation Systems (Airports, Seaport)
— Stuart D. Werner, *Dames & Moore*
C. Development of Real-Time Monitoring and Response Systems
— William Savage, *PG&E*

BRIDGES
— Ian Buckle, NCEER,
  Moderator
A. Steel (Bay Bridge)
— (to be announced),
  CALTRANS
B. Concrete (Cypress Viaduct)
— Jack P. Moehle, *Earthquake
  Engineering Research
  Center, University of
  California, Berkeley*
C. Performance of Codes and
  Retrofits
— Saiid Saiidi, *University of
  Nevada, Reno (invited)*

RECOVERY, MITIGATION
AND PLANNING
— Dennis Wenger, Texas
  A & M University,
  Moderator
A. Santa Cruz Historic District
— Richard Wilson, *Santa Cruz
  City Manager*
B. Housing
— Mary C. Comerio,
  *University of California,
  Berkeley*

C. Business Recovery
— Jack Kartez, *Texas A & M
  University*
D. Recovery in Two
  Communities
— Charles Eadie, *Watsonville
  Planning Department*

4:00 pm Closing Session
         "PRACTICAL LESSONS
         FROM THE LOMA PRIETA
         EARTHQUAKE: WHERE DO
         WE GO FROM HERE?"
         — Vitelmo V. Bertero,
           *University of California,
           Berkeley*

4:30 pm General Discussion
         Closing Remarks
         — Lloyd Cluff, *Chair, NRC
           Organizing Committee*

5:00 pm Adjourn

APPENDIX
C

Attendees at the Symposium

Manny Abad
Pacific Gas & Electric Co.
F1502A, One California St.
San Francisco, CA 94106

Massoud Abolhoda
City of Fremont
39550 Liberty Street
PO Box 5006
Fremont, CA 94537

Rafael Alaluf
H J Degenkolb Associates
350 Sansome St., Ste 900
San Francisco, CA 94104

Clarence R. Allen
Calif Inst of Technology
Seismological Lab 252-21
Pasadena, CA 91125

Eddy Alvarado
Citwin Engineers & Contractors
13551 Sharpbill Dr.
Houston, TX 77083

Brian Anderson
Snohomish Chiefs Assoc
16819 13th Avenue, W
Snohomish, WA 98037

Charles E. Anderson
U.S. Bureau of Reclamation
PO Box 25007D-3130
Denver, CO80225

Desideria Anderson
California State University
Geology Dept.
Northridge, CA 91330

William Anderson
National Science Foundation
1800 G Street, NW, Rm 1130
Washington, DC 20550

Richard Andrews
Calif Office of Emerg Services
2151 East D St., Ste 203A
Ontario, CA 91762

Matt Arabasz
University of Utah
705 W C Browning Bldg.
Salt Lake City, UT 84112

Walter Arabasz
University of Utah
705 W C Browning Bldg.
Salt Lake City, UT 84112

Christopher Arnold
Building Systems Development
3130 La Selva, Ste 308
San Mateo, CA 94403

Elizabeth Arscott
EERI Staff
499 14th Street, Ste 320
Oakland, CA 94612

Habte Asfaha
City of Richmond
PO Box 4046
Richmond, CA 94804

John R. Ashelin
Anheuser-Busch
2816 So 3rd St.
St Louis, MO 63005

Scott Ashford
University of California
Civil Engineering Dept.
Berkeley, CA 94720

Richard E. Austin
Fire Department
1234 Yates Street
Victoria BC Canada V8V 3M8

Otto Avvakumovits
GFDS Engineers
675 Davis Street
San Francisco, CA 94111

Bruce P. Baird
Calif OES-C S T I
PO Box 8104
San Luis Obispo, CA 93403

Ronald Bajuniemi
Harza Kaldveer
425 Roland Way
Oakland, CA 94621

William Bakun
US Geological Survey
345 Middlefield Rd, MS 977
Menlo Park, CA 94025

Donald Ballantyne
Kennedy/Jenks/Chilton
530 S 336th Street
Federal Way, WA 98003

Jerold H. Barnes
Salt Lake County Planning
2001 South State #N3700
Salt Lake City, UT 84190

Jim Battey
Smith-Emery Co.
PO Box 880550
Hunter's Pt Bldg.
San Francisco, CA 94588

James E. Beavers
Martin Marietta Energy Systems
PO Box 2009, Bldg 9207
Oak Ridge, TN 37831

Thomas R. Beckham
South Carolina Emrg Prep Div
1429 Senate St.
Columbia, SC 29201

James Bela
Oregon Earthquake Awareness
3412 SE 160th Avenue
Portland, OR 97236

Joseph W. Berg Jr.
3319 Dauphine Dr.
Falls Church, VA 22042

Lillian D. Berg
Northern Virginia Comm College
3319 Dauphine Dr.
Falls Church, VA 22042

Keith H. Bergman
Harding Lawson Associates
303 2nd St., Ste 630N
San Francisco, CA 94107

Vitelmo V. Bertero
University of California
783 Davis Hall
Berkeley, CA 94720

Elizabeth Z. Bialek
Roger Foott Associates
530 Howard St., 4th Floor
San Francisco, CA 94105

John W. Bitoff
Mayor's Office of Emerg Serv
1003A Turk St.
San Francisco, CA 94102

Robert Bolin
New Mexico State University
Box 300011, Dept 3BV
Las Cruces, NM 88003

Patricia A. Bolton
Battelle Human Affairs Res Ctr
4000 NE 41st Street
Seattle, WA 98105

David R. Bonneville
H. J. Degenkolb Associates
350 Sansome St., Ste 900
San Francisco, CA 94109

Roger D. Borcherdt
U.S. Geological Survey
345 Middlefield Rd, MS 977
Menlo Park, CA 94025

Ed Bortugno
Calif OES-EQ Program
101 8th St, Ste 152
Oakland, CA 94607

Linda Bourque
University of California
10833 Le Conte Ave.
Los Angeles, CA 90024

Jack G. Bouwkamp
Technical University Darmstadt
Alexanderstrasse 7
6100 Darmstadt Germany

George Bowman
Flames Training Service
PO Box 3657
Simi Valley, CA 93093

Gloria Bowman
Flames Training Service
PO Box 3657
Simi Valley, CA 93093

Gerald Brady
U.S. Geological Survey
345 Middlefield Rd.
Menlo Park, CA 94025

Robert Bridwell
Calif Dept of Transportation
PO Box 23660
Oakland, CA 94623

Theodor L. Brock
Navy Public Works Center
Oakland Army Base
Oakland, CA 94623

Ronald Brody
Naval Facilities Engr Command
900 Commodore Dr.
San Bruno, CA 94066

Roger Brown
Sear-Brown Group
1275 Columbus, #200
San Francisco, CA 94133

Ian G. Buckle
NCEER
Univ of Buffalo
105 Red Jacket Quad
Buffalo, NY 14261

James Buika
FEMA Region IX
Building 105, Presidio
San Francisco, CA 94129

Jane A. Bullock
FEMA-Natl EQ Hazards Reduction
500 C Street, SW
Washington, DC 20476

Tianqing Cao
SMIP
801 K Street, MS 13-35
Sacramento, CA 95814

Viet Cao
Pacific Gas & Electric Co.
F1502A, One California St.
San Francisco, CA 94106

Fortino Cardenas
Department of Housing
1800 3rd Street
Sacramento, CA 94253

Fred Caswell
The Rockland Co.
PO Box 24
Castro Valley, CA 94546

Donald Chan
Utilities Engineering Bureau
1155 Market St.
San Francisco, CA 94103

Andrew Charleston
School of Architecture
PO Box 600
Wellington, New Zealand

C. V. Chelapati
Calif State Univ-Long Beach
5912 Bolsa Ave., Ste 106
Huntington Beach, CA 92649

Hsiang Cheng
Pacific Gas & Electric Co.
F1502A, One California St.
San Francisco, CA 94106

Sue Chou
City & County of San Francisco
1680 Mission St.
San Francisco, CA 94103

Nadja Christian
OES/Coastal Region EQ Prog
101 8th St, Ste 152
Oakland, CA 94607

John H. Clark
Anderson Bjornstad Kane Jacobs
220 West Harrison Street
Seattle, WA 98119

Fred Clement
City of Richmond
PO Box 4046
Richmond, CA 94804

G. Wayne Clough
VPI & State University
College of Engineering
333 Norris Hall
Blacksburg, VA 24061

Lloyd S. Cluff
Pacific Gas & Electric
1 California St., Rm F-2200
PO Box 770000
San Francisco, CA 94177

Bud Coale
Novalert
1621 S Rancho Santa Fe Rd.
San Marcos, CA 92069

Kelly Cobeen
GFDS Engineers
675 Davis Street
San Francisco, CA 94111

David Cocke
H J Degenkolb Associates
350 Sansome St., Ste 900
San Francisco, CA 94104

Ronald M. Colas
CALTRANS
111 Grand Ave.
Oakland, CA 94623

Mary C. Comerio
University of California
232 Wurster Hall
Berkeley, CA 94720

Louise K. Comfort
University of Pittsburgh
Grad Sch of Pub & Intl Affairs
3E31 Forbes Quad
Pittsburgh, PA 15260

Guy J. Conversano
LACO Associates
216 J Street
Eureka, CA 95501

Richard Cook
U.S. Army Corps of Engineers
211 Main Street
San Francisco, CA 94105

James D. Cooper
Federal Highway Administration
6300 Georgetown Pike
McLean, VA 20164

Joseph M. Costello
Calif OES-C S T I
PO Box 8104
San Luis Obispo, CA 93403

William R. Cotton
William Cotton & Associates
330 Village Lane
Los Gatos, CA 95030

Tom Coull
Thomas B. Coull & Associates
654 Banyan Avenue
Walnut Creek, CA 94598

Chris H. Cramer
Division of Mines & Geology
801 K Street, MS 13-35
Sacramento, CA 95624

James H, Cullen
M Cullen Instruments
1090 Bodega Avenue
Petaluma, CA 94952

Donald A Cushing, Jr.
ALX Engineering
1019 Solano Ave., Ste B
Albany, CA 94706

Arthur Dang
OSA/SSS
301 Howard St., 4th Floor
San Francisco, CA 94105

Robert B. Darragh
SMIP
801 K Street, MS 13-35
Sacramento, CA 95814

Robert D. Darragh
Dames & Moore
221 Main Street
San Francisco, CA 94105

Jim Davidson
C & M Consultants
933 Canyon Blvd.
North Vancouver BC Canada V7R 2J9

John Davis
3716 West 28th Avenue
Vancouver, BC Canada

Noemi de Guzman
Kaiser Permanente-No Calif
1950 Franklin St, 12th Floor
Oakland, CA 94612

Veronica Dekovic
City of Fremont
39550 Liberty Street
PO Box 5006
Fremont, CA 94537

Don Delcourt
BC Hydro
12685 82nd Avenue
Surrey BC Canada V3W 3G2

Stephen Dickenson
Oregon State University
Civil Engineering Dept.
Apperson Hall 202
Corvallis, OR 97331

Cleto Dosremedios
BC Hydro
6911 Southpoint Dr., Podium A03
Burnaby, BC Canada

Bruce M. Douglas
University of Nevada
Civil Engineering Dept
Earthquake Research
Reno, NV 89557

Tom Durham
CUSEC
2630 E Holmes Rd.
Memphis, TN 38118

Michael E. Durkin
Michael E. Durkin & Associates
22955 Leonora Dr.
Woodland Hills, CA 91367

Andrew Dys
V I Territory Emergency Agency
2 C Contant
St Thomas, VI 00802

Charles Eadie
Planning Department
250 Main Street
Watsonville, CA 95076

Wayne D. Edwards
Pacific Gas & Electric
PO Box 770000, Mail Code F16A
San Francisco, CA 94177

Ronald Eguchi
EQE Engineering & Design
18101 Von Karman Ave.
Lake Shore Towers, Ste 400
Irvine, CA 92715

Richard Eisner
Calif OES-EQ Program
101 8th St, Ste 152
Oakland, CA 94607

Mark W. Eli
Lawrence Livermore Natl Lab
PO Box 808, MS L-194
Livermore, CA 94550

Robert Terry Elliott
State Farm Insurance
112 E Washington St.
Bloomington, IL 61701

William Ellsworth
U.S. Geological Survey
345 Middlefield Rd.
Menlo Park, CA 94025

James Esse
E W Blanch Co.
3500 W 80th Street
Minneapolis, MN 55431

Ernesto Eugenio
U.S. Army Corps of Engineers
211 Main Street
San Francisco, CA 94105

Stephen S. Fader
Engineering Division/Wellesley
455 Worcester St
Wellesley Hills, MA 02181

James R. Faris
NAVFAC-Geotech Branch
900 Commodore Dr.
San Bruno, CA 94066

Kent S. Ferre
Pacific Gas & Electric
PO Box 770000, Mail Code F16A
San Francisco, CA 94177

Sands Figuers
Rogers/Pacific
396 Civil Dr.
Pleasant Hill, CA 94528

James Finley
Hesselberg, Kessee & Assoc.
221 Main St., Ste 1580
San Franciso, CA 94105

Clement Finney
Diocese of Oakland
2900 Lakeshore Ave
Oakland, CA 94610

Catherine Firpo
Calif OES-EQ Program
101 8th St., Ste 152
Oakland, CA 94607

Albert E. Fischer III
Environmental Protection Agcy
75 Hawthorne St., H-1-2
San Francisco, CA 94105

Louis Flores
CALTRANS
10 E Vine St., Ste 214
Redlands, CA 92373

Eric Fok
Pacific Gas & Electric Co.
F1502A, One California St.
San Francisco, CA 94106

Marie L. Fong
Navy Public Works Center
Oakland Army Base
Oakland, CA 94623

Christine M. Forakis
Statewide Health Plan & Dev
1600 P St., Room 420
Sacramento, CA 95814

Nicholas F. Forell
Forell/Elsesser Engineers
539 Bryant St., 3rd Floor
San Francisco, CA 94107

Howard Foster
Univ of California-Berkeley
217 Monte Vista Avenue
Oakland, CA 94611

Julian F. Foster
USEPA, Region VI
1445 Ross Ave, 6E-EP
Dallas, TX 75202

Tom Foster
Snohomish Chiefs Assoc
16819 13th Avenue, W
Snohomish, WA 98037

Paul F. Fratessa
Paul A. Fratessa Associates
360 22nd Street
Oakland, CA 94612

Sigmund A. Freeman
Wiss, Janney, Elstner Assoc
2200 Powell St., Ste 925
Emeryville, CA 94608

Laurie Friedman
EQE International
44 Montgomery St., Ste 3200
San Francisco, CA 94115

Katie Frohmberg
EERC
1301 S 46th Street
Richmond, CA 94804

Fred Fung
Dept of Building Inspection
651 Pine St.
Martinez, CA 94553

Lind S. Gee
University of California
Seismological Station
475 Earth Sciences Bldg.
Berkeley, CA 94720

Carl A. Gentry
John Carollo Engineers
450 N Wiget Lane
Walnut Creek, CA 94598

William T. Gin
Navy Public Works Center
Oakland Army Base
Oakland, CA 94623

Larry Goldfarb
Harza Kaldveer
425 Roland Way
Oakland, CA 94621

James Goltz
Calif OES/EQ Prog/Southern Reg
111- E Green St, Ste 300
Pasadena, CA 91106

Soma Goresky
Pacific Geotechnical Engr.
16120 D Caputo Dr.
Morgan Hill, CA 95037

James P. Gould
Nueser Rutledge Cons Eng
708 3rd Ave
New York, NY 10017

Leslie W. Graham
Graham & Kellam
3 Crest Road
Belvedere, CA 94920

Jack Greene
Earth Systems
670 Darrell Rd.
Hillsborough, CA 94010

Marjorie Greene
OES/Coastal Region EQ Prog
101 8th St, Ste 152
Oakland, CA 94607

Joe Grindstaff
Eastern Municipal Water Dist
PO Box 8300
San Jacinto, CA 92383

Caroline Clarke Guarnizo
National Academy of Sciences
2101 Constitution Ave, NW
Washington, DC 20418

Paul Guerin
ENGEO Inc.
2401 Crow Canyon Rd, Ste 200
San Ramon, CA 94585

Luis Gutierrez
CALTRANS
1169 Market St., #356
San Francisco, CA 94103

Amanda Houston Hamilton
Polaris R & D
185 Berry St., Ste 6400
San Francisco, CA 94107

Thomas C. Hanks
U.S. Geological Survey
345 Middlefield Rd.
Menlo Park, CA 94025

Joyce E. Harris
L.A. County Emergency Services
500 W Temple St, Rm 783
Los Angeles, CA 90012

Marley Hart
Statewide Health Plan & Dev
1600 9th St., Room 420
Sacramento, CA 95814

Suzanne K. Hart
City/County of San Francisco
Public Works Dept.
2323 Army Street
San Francisco, CA 94124

Danny Hattaway
State Farm Insurance
112 East Washington St.
Bloomington, IL 61701

Bob Hayden
Hayward Baker
1780 Lemonwood
Santa Paula, CA 93060

Dean Haymore
Storey Co Building Official
PO Box 526
Virginia City, NV 89440

Walter W. Hays
U.S. Geological Survey
905 National Center
Reston, VA 22092

Patricia Heerhartz
BSB
1600 9th St., Room 420
Sacramento, CA 95814

A C Heidebrecht
McMaster University
1280 Main St. West
Hamilton, ON Canada

Sam Henderson
Dept. of Emergency Mgmt
1907 Everett Avenue
Everett, WA 98201

Michael E. Heppell
Fire Department
1234 Yates Street
Victoria BC Canada V8V 3M8

Francois E. Heuze
Lawrence Livermore Natl Lab
PO Box 808
Livermore, CA 94551

Steven T. Hiner
Cole/Yee/Schubert & Associates
2500 Venture Oaks Way, Ste 200
Sacramento, CA 95833

Ephraim G. Hirsch
E G Hirsch & Associates
Pier 1 1/2, The Embarcadero
San Francisco, CA 94111

A. J. Hitchings
EQE International
44 Montgomery St., Ste 3200
San Francisco, CA 94104

Joanne Hoffard
FEMA, Region IX
Bldg 105, The Presidio
San Francisco, CA 94129

Joe Hollstein
City of Ceres
PO Box 217
Ceres, CA 95307

William T. Holmes
Rutherford & Chekene
303 Second St., Ste 800 North
San Francisco, CA 94107

Thomas L. Holzer
US Geological Survey
345 Middlefield Rd.
Menlo Park, CA 94025

John C. Hom
John C. Hom & Associates
1618 2nd St
San Rafael, CA 97901

Willard Hopkins
Van Houten Consultants Inc.
422 Petaluma Blvd North
Petaluma, CA 94952

Terry Hoskins
H J Degenkolb Associates
350 Sansome St., Ste 900
San Francisco, CA 94104

Jan M. Hueser
HNTB Corp.
1001 Galaxy Way, Ste 101
Concord, CA 94520

Thom Huggett
Dept of Building Inspection
651 Pine St.
Martinez, CA 94553

Tom Hui
Utilities Engineering Bureau
1155 Market St.
San Francisco, CA 94103

Craig Huntington
Huntington Design Associates
4112 Park Blvd.
Oakland, CA 94602

Lawrence J. Hutchings
Lawrence Livermore Laboratory
PO Box 808, L-208
Livermore, CA 94551

Robert E. Isherwood
CIGNA
717 Appleberry Dr.
San Rafael, CA 94903

Wilfred D. Iwan
California Inst of Technology
Mail Code 104-44
Pasadena, CA 91125

Wayne C. Jarvis
United Methodist Relief Comm
502 Porter
Helena, AR 72342

Curtis Jensen
Jensen-Van Lienden Associates
1757 Alcatraz Avenue
Berkeley, CA 94703

Gary Johnson
FEMA
Off of EQ & Nat Hazards
500 C Street, SW
Washington, DC 20472

Gerald Jones
City of Kansas City
18th Floor, City Hall
Kansas City, MO 64106

Nicholas P. Jones
Johns Hopkins University
Civil Engineering Dept.
Baltimore, MD 21218

Chris Jonientz-Trisler
FEMA
130-228th Street, SW
Bothell, WA 98021

Carl Josephson
Carl Josephson & Associates
6640 Lusk Blvd, Ste A200
San Diego, CA 92121

Kenneth K. Kaestner
Kaestner Architects/Engineers
1600 Sunrise Ave., #3
Modesto, CA 95350

Raoul Karp
H J Degenkolb Associates
350 Sansome St., Ste 900
San Francisco, CA 94104

Jack D. Kartez
Texas A & M University
Hazard Reduction & Recovery
Langford Bldg C
College Station, TX 77843

Robert Katt
National Academy of Sciences
2101 Constitution Ave., NW
Washington, DC 20418

Dominic Kelly
H. J. Degenkolb Associates
350 Sansome St., Ste 900
San Francisco, CA 94104

Sami Kilic
Stanford University
Civil Engineering Dept
Stanford, CA94305

Stephanie A. King
Stanford University
John Blume EQ Res Ctr
Stanford, CA 94305

Charles Kircher
Kircher & Associates
444 Castro St, Ste 400
Mountain View, CA 94041

Anne Kiremidjian
Stanford University
Civil Engineering Dept
Terman 238
Stanford, CA 94305

Elaine Kissil
First Interstate Bank
1200 W 7th St, G8-10
Los Angeles, CA 90017

Mark Kluver
Portland Cement Association
83 Ryegate Pl.
San Ramon, CA 94583

Laurence Kornfield
Bureau of Building Inspection
1390 Market St, Ste 250
San Francisco, CA 94102

Herman Kwan
BC Hydro
6911 Southpoint Dr, Podium A03
Burnaby, BC Canada

Edmund Kwong
City of Fremont
39550 Liberty St.
PO Box 5006
Fremont, CA 94537

Libby Lafferty
Lafferty & Associates Inc.
4529 Angeles Crest Hwy, #308
PO Box 1026
La Canada, CA 91011

Henry J. Lagorio
10 Donald Dr.
Orinda, CA 94563

Wen J. Lai
Boeing
PO Box 3707, MS 17-MA
Seattle, WA 98124

Randolph Langenbach
FEMA, Region IX
Bldg 105, The Presidio
San Francisco, CA 94127

Susan Larsen
U.S. Geological Survey
345 Middlefield Rd, MS 977
Menlo Park, CA 94025

Patrick Lau
Utilities Engineering Bureau
1155 Market St.
San Francisco, CA 94103

Robert W. Lau
EBMUD
375 11th Street
Oakland, CA 94623

Robert Scott Lawson
Stanford University
649 Mirada Avenue
Stanford, CA 94805

Catherine K. Lee
General Services Admin.
525 Market St., 31st Flr
San Francisco, CA 94105

Christopher T. Lee
Navy Public Works Center
Oakland Army Base
Oakland, CA 94623

Kaiman Lee
Navy Public Works Center
Oakland Army Base
Oakland, CA 94623

Dawn Lehman
University of California
Civil Engineering Dept.
Berkeley, CA 94720

William Lettis
William Lettis & Associates
1000 Broadway, Ste 612
Oakland, CA 94607

Leo Levenson
FEMA, Region IX
Bldg 105, The Presidio, DAP
San Francisco, CA 94129

Tore Liahjell
DMJM
222 Kearny St., Ste 500
San Francisco, CA 94108

Joshua D. Lichterman
Emergency Management Group
1884 San Antonio Ave
Berkeley CA 94707

W. J. Lindblad
ACRS-USNRC
6770 SW Raleighwood Lane
Portland, OR 97225

Allan Lindh
U.S. Geological Survey
345 Middlefield Rd.
Menlo Park, CA 94025

Grant Lindley
University of California
Inst for Crustal Studies
Santa Barbara, CA 93106

Bret Lizundia
Rutherford & Chekene
303 2nd Street, #800N
San Francisco, CA 94107

David W. Look
National Park Service
600 Harrison St., Ste 600
San Francisco, CA 94107

Youzhi Ma
Geomatrix Consultants Inc.
100 Pine Street, 10th Floor
San Francisco, CA 94111

George Mader
William Spangle & Associates
3240 Alpine Rd.
Portola Valley, CA 94025

Janiele Tovani Maffei
H J Degenkolb Associates
350 Sansome St., Ste 900
San Francisco, CA 94104

Stephen A. Mahin
University of California
777 Davis Hall
Berkeley, CA 94720

Praveen Malhotra
Strong Motion Inst Prog
801 K St, MS 13-35
Sacramento, CA 95814

James E. Marrone
Bechtel Corp
PO Box 193965 MS 45/31
San Francisco, CA 94119

Steven R. Marten
DAS/Emergency Management
2320 4th Avenue
Seattle, WA 98121

James R. Martin II
VPI & State University
111B Patton Hall
Blacksburg, VA 24061

M. Lyn Martin
California Casualty
PO Box M
San Mateo, CA 94402

Roger Martin
Division of Mines & Geology
801 K Street, MS 32-12
Sacramento, CA 95814

Thomas Maruyama Jr.
401 Marshall St.
Redwood City, CA 94063

Ed Matsuda
Pacific Gas & Elctric Co.
One California St., Rm F-2200
PO Box 770000
San Francisco, CA 94177

Shirley Mattingly
Office of Emergency Management
200 N Main St, Rm 300
Los Angeles, CA 90012

Brent A. Maxfield
LDS Church
50 E North Temple
Salt Lake City, UT 84150

Sergio Mayorga
Pacific Gas & Electric Co.
201 Mission, Rm 841
San Francisco, CA 94117

Steven McAdam
S.F. Bay Conserv & Develop Comm
30 Van Ness Ave, Ste 2011
San Francisco, CA 94102

Richard McCarthy
Seismic Safety Commission
1900 K Street, #100
Sacramento, CA 95314

Frank E. McClure
Consulting Structural Engineer
54 Sleepy Hollow Lane
Orinda, CA 94563

Timothy McCrink
Division of Mines & Geology
801 K Street, MS 12-31
Sacramento, CA 95814

Robin K. McGuire
Risk Engineering Inc.
5255 Pine Ridge Rd.
Golden, CO 80403

Marcia K. McLaren
Pacific Gas & Electric Co.
One California St., Rm F-2200
PO Box 770000
San Francisco, CA 94177

R. Wayne McLeod
BC Hydro
12685 82nd Avenue
Surrey BC Canada V3W 3G2

Kurt McMullin
University of California
Civil Engineering Dept.
Berkeley, CA 94720

William M. Medigovich
FEMA Region IX
Building 105, Presidio
San Francisco, CA 94129

Daniel Meyerson
Cornell University
Civil Engineering Dept
Ithaca, NY 14853

Mike Michalski
City of Los Angeles
2426 Actman St.
Los Angeles, CA 90031

David Middleton
Earthquake Commission
PO Box 31-342
20 Daly St
Lower Hutt, New Zealand

John Miller
SCA
1104 W Airport Blvd, Ste 222
Stafford, TX 77477

John A. Miller
Nolte & Associates
60 S Market Street, Ste 300
San Jose, CA 95113

Marie L Minghini
High-Point
201 Spear St., Ste 1620
San Francisco, CA 94105

Masamitsu Miyamara
Kajima Corp
K-I Bldg
6-5-39 Akasaka, Minato-ku
Toyko, Japan

Kaoru Mizukoshi
Kajima Corp
19-1, Tobitskyu 2-chome
Chofu-shi
Toyko, 182 Japan

Jack P, Moehle
EERC-University of California
1301 South 46th Street
Richmond, CA 94804

Jimmy Moore
Memphis City Council
125 North Main, Ste 514
Memphis, TN 38103

Ugo Morelli
FEMA
Off of Nat & Tech Hazards
500 C Street, SW
Washington, DC 20472

Terry Morrow
City of Tacoma
Public Works Dept.
747 Market St., Ste 345
Tacoma, WA 98402

Carl E. Mortensen
U.S. Geological Survey
345 Middlefield Rd, MS97
Menlo Park, CA 94024

James Mullen
DAS/Emergency Management
2320 4th Avenue
Seattle, WA 98121

Seiichi Muramatsu
Kajima Corp
6-5-30 Akasaka Minato-ku
Toyko, Japan

Jay P. Murphy
Murphy Pacific Corp.
5630 Margarido Dr.
Oakland, CA 94618

Lynn Murphy
OES/Coastal Region EQ Program
101 8th St., Ste 152
Oakland, CA 94607

Randy L. Muth
University of Alaska
Duckering Bldg.
Fairbanks, AK 99775

Ernest Naesgaard
Macleod Geotechnical
1451 Marine Dr, Ste G
West Vancouver, BC Canada

Sarah K. Nathe
OES/Coastal Region EQ Program
101 8th St., Ste 152
Oakland, CA 94607

Dru R. Nielson
DCM/Joyal Engineering
484 N Wiget Lane
Walnut Creek, CA 94598

Dennis N. Nishikawa
Board of Public Works
200 N Spring St.
City Hall, Rm 370
Los Angeles, CA 90012

Ray Noble
Salt Lake County
2001 South State St, N3500
Salt Lake City, UT 84115

Camillo Nuti
Univ G D'Annunzio-Pescara
Viale Pindaro 42
65100 Pescara, Italy

Susan Nyman
Van Houten Consultants Inc
422 Petaluma Blvd, North
Petaluma, CA 94952

Naoto Ohbo
Kajima Corp
19-1 Tobitakyu
2-chome, Chofu-shi
Toyko, Japan

Jim Onderka
H J Degenkolb Associates
350 Sansome St., Ste 900
San Francisco, CA 94104

Thomas D. O'Rourke
Cornell University
265 Hollister Hall
Ithaca, NY 14853

Gregory Orsolini
Parsons-DeLeuw, Inc.
120 Howard Street
PO Box 3821
San Francisco, CA 94119

Keith Ory
Johnson &. Higgins
345 California Street
San Francisco, CA 94104

C. (Jack) Ouyang
3M Corporation
PO Box 33331/Bldg 42-8E-04
St Paul, MN 55133

Robert Packard
City of Los Angeles
600 So Spring St, Rm 1600
Los Angeles, CA 90014

Jorge R. Palafox
San Francisco EMS Agency
1540 Market St., Ste 220
San Francisco, CA 94104

Al Panico
American Red Cross
1660 Amphlett Blvd., Ste 312
San Mateo, CA 94402

Pervez Patel
Bureau of Building Inspection
450 McAllister St, Room 203
San Francisco, CA 94102

James Paustian
H. J. Degenkolb Associates
350 Sansome St., Ste 900
San Francisco, CA 94104

Laurence Pearce
Emergency Preparedness Canada
PO Box 10000
Victoria, BC Canada

Laurie Pearce
Univ of British Columbia
Disaster Preparedness Res Ctr
Vancouver BC Canada V6T 1Z3

Jonathan Pease
Cornell University
Civil Engineering Dept
Ithaca, NY 14853

Patricia Peck
EERI Staff
499 14th Street, Ste 320
Oakland, CA 94612

Gary Pedersen
City of Tacoma
Public Works Dept.
747 Market St., Ste 345
Tacoma, WA 98402

John Pender
Eastern Municipal Water Dist.
PO Box 8300
San Jacinto, CA 92581

Vern Persson
Dept. of Water Resources
2200 X Street
PO Box 942836
Sacramento, CA 94236

Pat Peterson
Dept of Building Inspection
451 So State St.
Salt Lake City, UT 84111

Steven H. Phillips
Pacific Gas & Electric Co.
One California St., Rm F-2200
PO Box 770000
San Francisco, CA 94177

Gary Pischke
Consultant
898 Union Street
Alameda, CA 94501

Chris D. Poland
H. J. Degenkolb Associates
350 Sansome St., Ste 900
San Francisco, CA 94104

Maurice S. Power
Geomatrix Consultants
100 Pine St., Ste 1000
San Francisco, CA 94111

Uday Prasad
City & County of San Francisco
1680 Mission Street
San Francisco, CA 94103

Gretchen Rau
University of California
Civil Engineering Dept.
Berkeley, CA 94720

Henry R. Renteria
City of Oakland
475-14th St., 9th Fl
Oakland, CA 94612

G. Michael Richards
Town of Woodside
PO Box 620005
Woodside, CA 94062

Mark R. Richards
Herrick & Richards
1684 Willamette St
Eugene, OR 97401

Charles Roberts
Port of Oakland
909 Ferry Street
Oakland, CA 94607

James G. Roberts
Calif Dept. of Transportation
PO Box 942874
Sacramento, CA 94274

Jacob Rodriguez
American Red Cross
1660 Amphlett Blvd., Ste 312
San Mateo, CA 94402

Don Rogers
A & B Engineers
7116 Saroni Drive
Oakland, CA 94611

Christopher Rojahn
Applied Technology Council
555 Twin Dolphin Dr, Ste 270
Redwood City, CA 94065

Richard D. Ross
Missouri St. Emrg Mgmt Agency
PO Box 116
Jefferson City, MO 65102

Richard J. Roth Jr.
California Insurance Dept.
18896 Cordata St.
Fountain Valley, CA 92708

Glen Roycroft
Miller Pacific Engr Group
165 N Redwood Dr, Ste 165
San Rafael, CA 94903

Lisa Russell
University of California
10833 Le Conte Ave.
Los Angeles, CA 90024

Mark R. Russo
FEMA-Natl EQ Hazards Reduction
500 C Street, SW
Washington, DC 20476

M. "Saiid" Saiidi
University of Nevada
Civil Engineering Dept MS 258
Reno, NV 89557

Cherel Sampson
City of Sunnyvale
700 All American Way
Sunnyvale, CA 94086

Laura A. Samrad
CALTRANS
111 Grand Avenue
Oakland, CA 94623

Gary J. Sander
HNTB Corporation
1001 Galaxy Way, Ste 101
Concord, CA 94520

Lee J Sapaden
Calif Dept of Social Serv
Disaster Response Serv
744 P St, MS 19-43
Sacramento, CA 95814

Sarwidi
Rensselaer Polytechnic Inst
Civil Engineer Dept
4037 JEC-RPI
Troy, NY 12180

William U. Savage
Pacific Gas & Electric
1 California St, Rm F-2200
PO Box 770000
San Francisco, CA 94177

Charles R. Scawthorn
EQE International
44 Montgomery St., Ste 3200
San Francisco, CA 94104

Stanley G. Schaffer
Anheuser-Busch
2816 So 3rd St.
St Louis, MO 63005

Tim Scheckler
Chevron Research & Technology
PO Box 1627
Richmond, CA 94802

Dove Sholom Scherr
Joshua B. Kardon & Co.
1939 Addison St., Ste A
Berkeley, CA 94704

Mark T. Schexnayder
Pacific Gas & Electric Co.
PO Box 770000, Mail Code F16A
San Francisco, CA 94177

Richard Schroedel
Pierce Co. Dept. of Emerg Serv
930 Tacoma Ave So., Rm #B-36
Tacoma, WA 98402

Diane Schurr
City of Tacoma
Public Works Dept.
747 Market St, Ste 345
Tacoma, WA 98402

Stanley Scott
University of California
1141 Vallecito Ct.
Lafayette, CA 94549

Sandra Seale
University of California
Inst for Crustal Studies
Santa Barbara, CA 93106

Linda C. Seekus
U.S. Geological Survey
345 Middlefield Rd, MS977
Menlo Park, CA 94025

Guna Selvaduray
San Jose State University
14829 Rossmoyne Drive
San Jose, CA 95124

Pasan Seneviratna
Stanford University
Civil Engineering Dept.
Stanford, CA 94305

Rocco Serrato
City of Culver City
4095 Overland Ave.
Culver City, CA 90232

Roland Sharpe
Consulting Structural Engineer
10320 Rolly Rd.
Los Altos, CA 94024

Subhash Shastri
Utilities Engineering Bureau
1155 Market St.
San Francisco, CA 94103

Robin Shepherd
University of California
Civil Engineering Dept.
Irvine, CA 92717

Craig Silvey
City of Tacoma
Public Works Dept.
747 Market St., Ste 345
Tacoma, WA 98402

Ajay Singhal
Stanford University
Civil Engineering Dept
Stanford, CA 94305

Nick Sitar
University of California
440 Davis Hall
Berkeley, CA 94720

Peter H. Smeallie
National Academy of Sciences
2101 Constitution Ave, NW
Washington, DC 20418

Paul Sommerville
Woodward-Clyde Consultants
566 El Dorado St.
Pasadena, CA 91101

Pam Soper
FEMA-Region VII
911 Walnut, Rm 200
Kansas City, MO 64106

Thomas Statton
Woodward-Clyde Consultants
101 Convention Ctr Dr, #P110
Las Vegas, NV 89109

Karl V. Steinbrugge
6851 Cutting Blvd.
El Cerrito, CA 94530

Bozidar Stojadinovic
University of California
Civil Engineering Dept.
Berkeley, CA 94720

Carl L. Strand
Stand Earthquake Consultants
1436 S Bentley Ave, #6
Los Angeles, CA 90025

Eric Straser
Stanford University
Civil Engineering Dept.
Stanford, CA 94305

Charles B. Stuart
FEMA-Disaster Assistance Ofc
500 C Street, SW
Washington, DC 20476

Paul Stukas
Rutherford & Chekene
303 2nd St., Ste 800 North
San Francisco, CA 94107

Lacey Suiter
Tennessee Emrg Mgmt Agcy
3041 Sidco Dr.
National Guard Armory
Nashville, TN 37204

Robert J. Swain
OYO Geospace Corp.
35 N Lake Ave.
Pasadena, CA 91101

Louie Tan
City of Salinas
200 Lincoln Ave.
Salinas, CA 93901

Cheryl Tateishi
California OES-EQ Program
1110 E Green St, Ste 300
Pasadena, CA 91106

Charles L. Taylor
Geomatrix Consultants
100 Pine St., Ste 1000
San Francisco, CA 94111

Nicoluos Theophanous
City of Concord
1950 Parkside Dr.
Concord, CA 94519

Barbara L. Thurlow
Pacific Gas & Electric Co.
One California St., Rm F-2200
PO Box 770000
San Francisco, CA 94177

Kathleen Tierney
University of Delaware
Disaster Research Center
Newark, DE 19716

John C. Tinsley
U.S. Geological Survey
345 Middlefield Rd.
Menlo Park, CA 94025

Tom Tobin
Calif Seismic Safety Comm
1900 K Street, Ste 100
Sacramento, CA 95814

Diana Todd
Natl Inst of Standards/Tech
Bldg. 226, Rm. B158
Gaithersburg, MD 20899

Ernie Tom
Utilities Engineering Bureau
1155 Market St.
San Francisco, CA 94103

Kenneth C. Topping
Urban Planning Consultant
1196 Banyan St.
Pasadena, CA 91103

Tousson Toppozada
Division of Mines & Geology
801 K Street, MS 12-31
Sacramento, CA 95814

Susan Tubbesing
Executive Director, EERI
499 14th St., Ste 320
Oakland, CA 94612

Fred Turner
Calif Seismic Safety Comm
1900 K Street, #100
Sacramento, CA 95814

Yi-Ben Tsai
Pacific Gas & Electric Co.
One California St., Rm F-2200
PO Box 770000
San Francisco, CA 94177

Randall G. Updike
U.S. Geological Survey
905 National Center
Reston, VA 22092

Gary E. Van Houten
Van Houten Consultants Inc.
422 Petaluma Blvd, North
Petaluma, CA 94952

Geoffrey Van Lienden
Jensen-Van Lienden Associates
1757 Alcatraz Avenue
Berkeley, CA 94703

David L. Vargo
American Red Cross
FEMA Region IX-Bldg 105
Presidio
San Francisco, CA 94129

Vasiliki Vassil
Smith-Evernden Associates
PO Box 1411
Capitola CA 95010

Dennis W. Vidmar
City of Bellevue
2901 115th Avenue, NE
Bellevue, WA 98004

Sara Wadia
Stanford University
Civil Engineering Dept.
Stanford, CA 94305

Jaunell J Waldo
Santa Clara County
962 Emory St
San Jose CA 95126

James A. Walke
FEMA-Disaster Assistance Ofc
500 C Street, SW
Washington, DC 20476

Yat Sing Wan
Navy Public Works Center
Oakland Army Base
Oakland, CA 94623

W. K. Warnock
Sverdrup Corporation
1340 Treat Blvd., Ste 100
Walnut Creek, CA 94596

Walter K Weibezahn
VHS Associates
1050 Northgate Dr
San Rafael CA 94903

Dennis E. Wenger
Texas A & M University
Hazard Reduction & Recovery
College Station, TX 77843

Stu Werner
Dames & Moore
2101 Webster St., Ste 300
Oakland, CA 94612

Robert Wesson
U.S. Geological Survey
905 National Center
Reston, VA 22092

Gael Wheeler
FEMA
911 Walnut
Kansas City, MO 64106

Gail Wiggett
Dept of Conservation
Division of Mines & Geology
801 K Street, MS12-32
Sacramento CA 95814

William B. Wigginton
Geolex, Inc.
PO Box 31151
Walnut Creek, CA 94598

Dale G. Wilder
Lawrence Livermore Natl Lab
7000 East Ave, PO Box 808
Livermore, CA 94550

Daryl Willey
City of Modesto
801 11th St, PO Box 642
Modesto CA 95353

Don F. Willoughby
Pacific Gas & Electric
PO Box 770000, Mail Code F16A
San Francisco, CA 94177

Chris Wills
Division of Mines & Geology
1145 Market St., 3rd Fl.
San Francisco, CA 94103

Richard Wilson
City of Santa Cruz
809 Center St
Santa Cruz CA 95060

Scott Wilson
University of California
EHS 0920
La Jolla, CA 92064

Frannie Winslow
City of San Jose-OES
855 N San Pedro St, #404
San Jose, CA 95110

Dean Wolf
Navy Public Works Center
PO Box 24003
Oakland, CA 94623

Wilda C. Wolf
Texas Instruments
6550 Chase Oaks Blvd.
PO Box 869305 MS 8475
Plano, TX 75086

Ivan G. Wong
Woodward-Clyde
500 12th Street
Oakland, CA 94607

Terry Wong
City of Los Angeles
650 S Spring St, Ste 400
Los Angeles, CA 90014

Ned Worcester
Group Health Corp
521 Wall Street
Seattle, WA 98121

Christopher Wright
Warner Brothers
4000 Warner Blvd, #34
Burbank, CA 91522

Richard Wright
Natl Inst of Standards/Tech
Bldg & Fire Lab
Bldg 226, Rm B216
Gaithersburg, MD 20899

Loring Wyllie
H J Degenkolb Associates
350 Sansome St., #900
San Francisco, CA 94104

T. Leslie Youd
Brigham Young University
Civil Engineering Dept-368 CB
Provo, UT 84602

Fan Yuan
Pacific Gas & Electric
PO Box 770000, Mail Code F16A
San Francisco. CA 94177

Susanne Zechiel
Michael Brandman Associates
606 S Olive St, Ste 600
Los Angeles, CA 90014

Domenic A. Zigant
NAVFAC
900 Commodore Dr.
San Bruno, CA 94066

Biographical Sketches of the Members of the Committee on Practical Lessons from the Loma Prieta Earthquake

Lloyd C. Cluff, Chair, is Manager of the Geosciences Department at the Pacific Gas and Electric Company. He is a member of the National Academy of Engineering. Mr. Cluff's areas of expertise are in seismic and geologic hazards, seismic geology, seismicity and paleoseismicity, engineering geology, environmental geology, and earthquake engineering. His research and experience include studies of the relationship of tectonics, seismic geology, and seismicity in many tectonic environments throughout the world. Mr. Cluff has completed fault activity and earthquake studies of many fault systems including those in New Zealand, Australia, Chile, Peru, Argentina, Ecuador, Bolivia, Colombia, Venezuela, Costa Rica, Nicaragua, Honduras, El Salvador, Guatemala, Mexico, Japan, Taiwan, India, Nepal, Pakistan, Iran, Afghanistan, Turkey, Egypt, Algeria, Russia, Israel, Jordan, Italy, Romania, Switzerland, Spain, and Portugal. He has served as an advisor to the governments of many of these countries regarding earthquake and geologic hazards and the formulation of seismic safety guidelines and public policy, especially in the siting and design of critical facilities. Mr. Cluff received his B.S. from the University of Utah.

Clarence R. Allen is Professor Emeritus of Geology and Geophysics at the California Institute of Technology. He is a member of the National Academy of Sciences and the National Academy of Engineering as well as a member of the American Association for the Advancement of Science and the American Association of Petroleum Geologists, a Fellow in the American Geophysical Union, and a Fellow of the Geological Society of America. Dr. Allen is past president of the Seismological Society of America and past president of the Geological

Society of America. His primary research interests include seismotectonics, tectonics of regional fault systems, earthquake prediction, and nuclear waste management. In 1989, Dr. Allen was appointed by the President to the Nuclear Waste Technical Review Board. He received his B.A. from Reed College and his M.S. and Ph.D. degrees from the California Institute of Technology.

Thomas R. Beckham has been employed by the South Carolina Emergency Preparedness Division since 1971. Since that time, he has held positions that include: Assistant Community Shelter Planning Officer, State Property Officer, Assistant Director of the Nuclear Branch, Chief Fixed Nuclear Facility Planner, and Operations Branch Manager. On July 1, 1986, he was appointed Deputy Director of the Emergency Preparedness Division and is currently serving in that position. Mr. Beckham serves on the National Coordinating Council on Emergency Management, the South Carolina Emergency Preparedness Association, the South Carolina Hurricane Task Force, and the South Carolina Wildland Fire Protection Partnership. Mr. Beckham represents an East Coast user of information from the Loma Prieta earthquake symposium.

Ian G. Buckle is Professor of Civil Engineering and Deputy Director of the National Center for Earthquake Engineering Research at the State University of New York at Buffalo. Prior to this, he was Director of Research and Development at Computech Engineering Services, Berkeley, and Vice President Engineering, Dynamic Isolation Systems. Dr. Buckle received his B.E. and Ph.D. degrees from the University of Auckland, New Zealand.

Wilfred (Bill) D. Iwan is Professor of Civil Engineering and Applied Mechanics at the California Institute of Technology. He is currently a member of the Board on Natural Disasters and chair of the Committee on Hazards Mitigation Engineering, and has chaired the former Committee on Natural Disasters, as well as served on the NRC's Committee on Earthquake Engineering, the California Seismic Safety Commission, the International Association for Earthquake Engineering, and many others. Dr. Iwan received his B.A., M.A., and Ph.D. from the California Institute of Technology.

Shirley Mattingly is Director of Emergency Management and Chief Administrative Analyst for the city of Los Angeles. She also serves as chairperson for the Emergency Preparedness Commission for the county and city of Los Angeles, vice-chairperson for the Policy Advisory Board of the Southern California Earthquake Preparedness Project, and vice-president of the Executive Committee of the Business and Industry Council for Emergency Planning and Preparedness. Ms. Mattingly is a current member of the Board on Natural Disasters. She received her B.A. from Occidental College, Los Angeles, and her M.A. from the University of California, Los Angeles.

Robin K. McGuire is President of Risk Engineering, Inc. His areas of expertise include earthquake engineering, risk analysis, decision analysis, and geostatistics. His experience includes directing a major project for Electric Power Research Institute to develop and apply methods of evaluating earthquake hazards for nuclear power plants in the eastern United States, consultant to the national Committee on Property Insurance regarding earthquake matters and recommendations to the California Department of Insurance, lead consultant on probabilistic performance assessment of the Yucca Mountain site as a possible high-level waste repository, and lead consultant on numerous studies of seismic and environmental risk for utilities, insurance groups, and commercial clients. Dr. McGuire received his S.B. in civil engineering from the Massachusetts Institute of Technology; M.S. in structural engineering from the University of California, Berkeley; and his Ph.D. in structural engineering from the Massachusetts Institute of Technology.

Chris D. Poland is President of H.J. Degenkolb Associates. His career has included a broad range of structural engineering projects, seismic evaluation, strengthening of existing buildings, and failure analysis work. Mr. Poland received his B.S. from the University of Redlands and his M.S. from Stanford University.

Dennis E. Wenger is Director, Hazard Reduction and Recovery Center, College of Architecture; Professor of Urban and Regional Planning; and Adjunct Professor of Sociology, all at the Texas A&M University. He was Co-Director of the Disaster Research Center at the University of Delaware from 1985 through 1989. Dr. Wenger was a staff member of the President's Commission on the Accident at Three Mile Island in 1979. His areas of expertise are in collective behavior, sociology of disaster and collective stress, sociology of mass communication, urban sociology, community power and social conflict, and research methodology. Dr. Wenger received his B.S., M.A., and Ph.D. from the Ohio State University.

T. Leslie Youd is Professor of Civil Engineering at Brigham Young University. Prior to his professorship he was a research and civil engineer at the U.S. Geological Survey. In 1982 Dr. Youd received a Superior Service Award from the Department of the Interior. He received his B.E.S. from Brigham Young University and his Ph.D. from Iowa State University.